UNDERSTANDING
ECLIPSES

CLIFF TURK

Struik Publishers (Pty) Ltd
(a member of the Struik New Holland Publishing Group (Pty) Ltd)
Cornelis Struik House
80 McKenzie Street
Cape Town 8001

First published in 2001

10 9 8 7 6 5 4 3 2 1

Publishing manager: Pippa Parker
Editor: Helen de Villiers
Designer: Dominic Robson
Design assistant: Illana Fridkin
Illustrator: Steven Felmore

Printed and bound by The Rustica Press Ndabeni, Western Cape
D8652

ISBN 1 86872 580 4

Cover photographs: (main picture) International Press Agency; (inset pictures) Cliff Turk
Title page photograph: *90% Solar Eclipse, Cape Town, 30 June 1999*: Cliff Turk

Contents

Acknowledgements

In writing this book I have been fortunate in having a circle of friends who have been full of encouragement and helpful suggestions. Some of them may not even realise how much time and effort they saved me. Without them, my task would have been much harder and I thank them all most sincerely. They are John Caldwell, Ilona Dicker, André Erasmus, Ian Glass, Tony Jones, Audrey Joubert, Ethleen Lastovica, Herschell Mair, Francis Podmore and Greg Roberts. If I have left anyone out, I apologise. Finally, if there are any mistakes in this book, these are nothing to do with my friends – the mistakes are all mine.

Introduction

No matter how many times one has seen it before, a total eclipse of the Sun is the most amazing experience. At the critical moment, when the Sun is completely obscured by the Moon, so great is the impact that viewers usually break into cheers. The difference between the partial phases, or a partial eclipse, and totality itself is like the proverbial chalk and cheese. If you have only ever seen partial eclipses, even if you were interested enough to use specialized telescopic or photographic equipment to record them, you cannot imagine the feeling of wonderment when the Moon completely blots out the Sun in the sky. Observers have been taken by surprise by the effect and there are many stories about people forgetting to take photographs or to look at the shadow racing across the countryside, while they were just gazing at the spectacle.

Southern Africa is to be the arena for two upcoming eclipses – one in June 2001, and one in December 2002. One of the most important things to consider about these eclipses is the site from which you will observe events. In both instances, totality will cover only a narrow strip across the continent, but a partial eclipse will be visible from any part of Africa that is south of the Equator, and in places even up to 20° north. (See maps 1 and 2 on pages 6 and 7.) However, the prevailing weather will obviously affect viewing from any position.

In Cape Town, for instance, the June 2001 eclipse should show 51% of the Sun's diameter covered by the Moon, but this is in a winter-rainfall area and there is quite likely to be a howling north-west wind with thick nimbus clouds and rain. If that happens, nothing will be seen. It will not even become dark, as Cape Town is far from the path of totality. Johannesburg should experience the reverse, as it usually has clear winter skies in June, for a 73% eclipse.

On the other hand, December 2002 should produce good weather in the Cape for the 60% eclipse – but Johannesburg, where the eclipse will be approximately 88%, may get rain and cloud, except, hopefully, in the early morning when the eclipse is due.

If you are planning to travel to view the eclipses, another consideration is the equipment to be taken along, especially if it is heavy and an air flight is involved. Contrary to popular belief, a large telescope is not necessary, but even telephoto lenses for cameras can be quite heavy. Optical and photographic equipment will be dealt with in more detail in the final chapter.

Perhaps the major problem will be where to stay, at or near your chosen observation site. There are likely to be many thousands of overseas visitors coming especially to see the eclipses and all holiday accommodation will rapidly be taken up. Problems might arise because of the prevailing relatively unstable political situation over much of the path of totality, with bookings needing to be made 6 to 12 months before the event.

It is generally best to arrive at your chosen venue a day or two in advance to allow at least one whole day to get your bearings, and to pick your observing position before the day of the eclipse. Make sure there is plenty of space. For the 11 August 1999 eclipse, the author chose the grounds in front of the War Memorial in Szombathely, Hungary. Next day, he arrived three hours before totality – and found about 200 people there before him. Fortunately there was ample space.

Lunar eclipses are much less dramatic than solar eclipses, but they nevertheless provide a spectacle that is well worth watching. There is no need to travel to any special area to see them, as they are visible from any part of the Earth which is in darkness at the right time – anyone who can see the Moon can see the eclipse.

Probably the most impressive thing about lunar eclipses is the frequent appearance of an orange colour caused by the refraction and scattering of light through the atmosphere of the Earth, although this colour is not always present to the same degree.

This book explains why and how eclipses occur, both solar and lunar. It deals principally, however, with the two upcoming solar eclipses that will be visible from large parts of southern Africa. Whether you wish to observe the eclipses seriously, or you prefer simply to enjoy the spectacle, this book is designed to help you get the greatest pleasure from your experience.

Plate 1: This photograph of the corona was taken in Szombathely, Hungary, during the solar eclipse of 1999. Inset are, from the left, Eric Dicker, Chris Turk (the author's son) and the author, Cliff Turk. They are viewing the eclipse with the use of commercially produced protective glasses.

Photo: Ilona Dicker

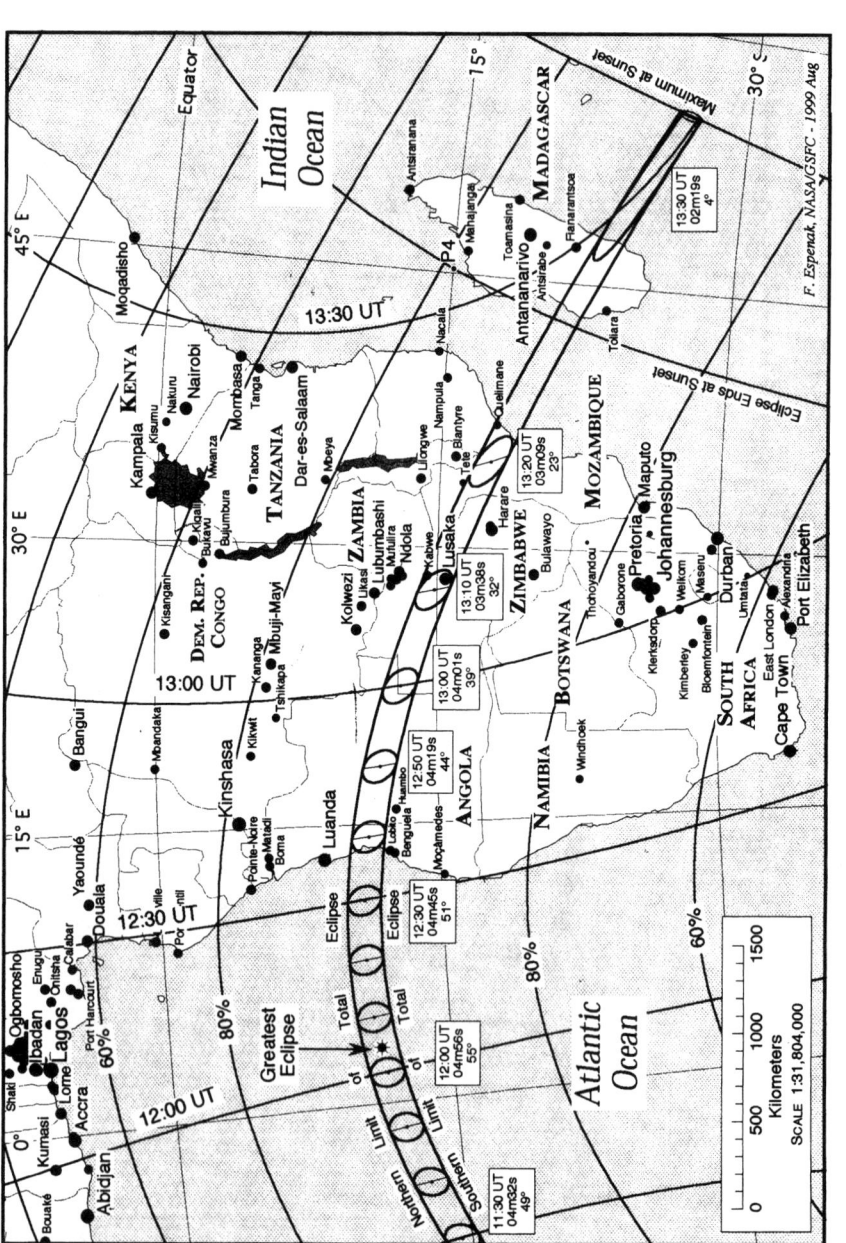

Map 1: Total solar eclipse of 21 June 2001 – the eclipse path through Africa.

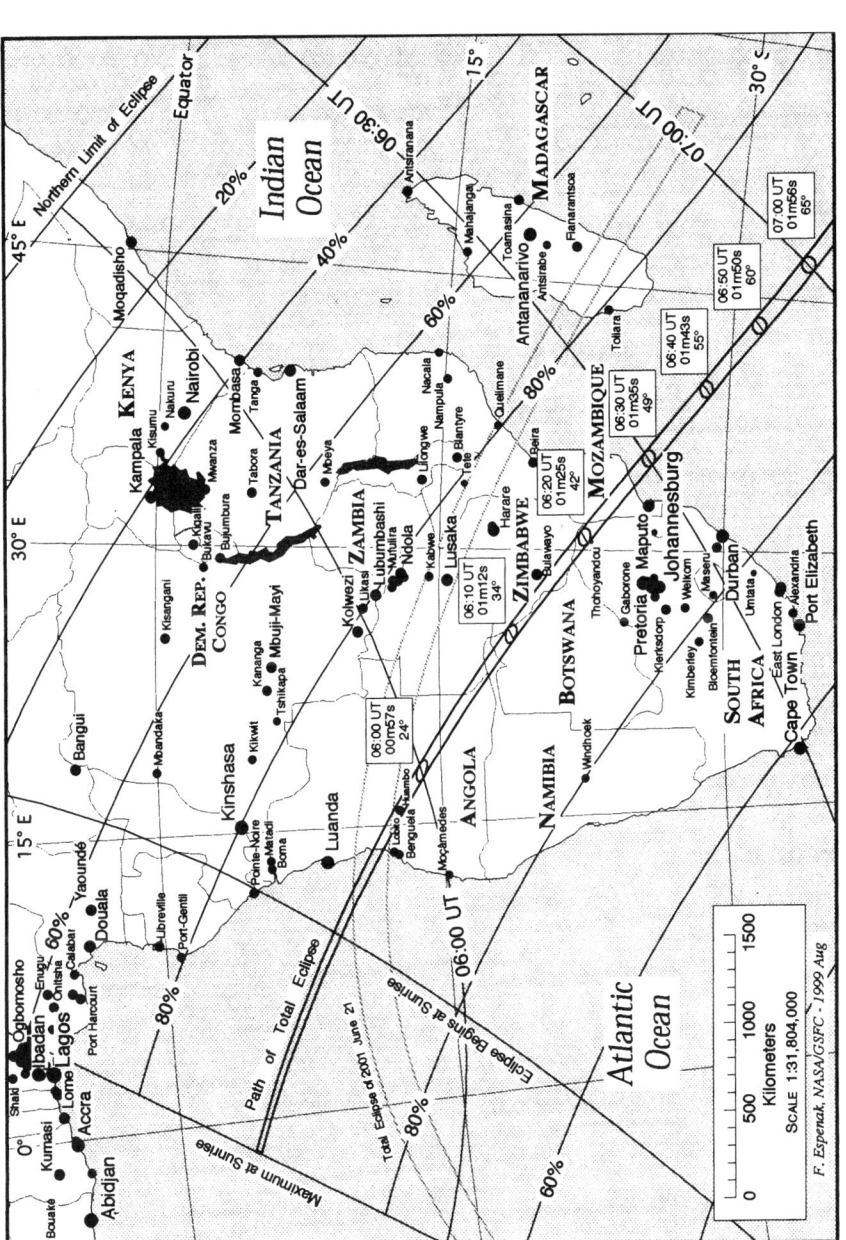

Map 2: Total solar eclipse of 4 December 2002 – the eclipse path through Africa.

Eclipses through history

The spectacle of a solar eclipse has always excited man's imagination. In mythology and in cultural records, eclipses were often seen as symbols of annihilation and doom, and the overthrow of the natural order of things. The word 'eclipse' comes from Latin and Greek words meaning 'to abandon'.

In many cultures, an eclipse was represented by the image of a dragon or demon that devoured the Sun. The ancient Chinese would frighten away the dragon by banging on pots and drums during an eclipse, creating a great noise and commotion. The Incas used similar methods to intimidate the beasts that were eating the Sun. In India, people would immerse themselves up to the neck in water to help fight off the dragon.

Written reports of past expeditions to observe solar eclipses give an idea of early responses to the phenomenon. During a total eclipse in 1859, a team of scientists watching from a position high up in the Andes noted the reactions of two local villagers '. . . watching in silence, and with anxious countenances, the rapid and fearful decrease of light. . . . At that instant (of totality), one exclaimed, in terror, 'La Gloria!' and both, I believe, fell to their knees in awe. They appreciated the resemblance of the corona to the halos . . . of our Savior and the Madonna, and devoutly regarded this as a manifestation of the divine presence.'

Detailed notes of the duties carried out by participants on expeditions, some in difficult locations, can be amusing. Timing, in particular, was a problem and every person in the team had his or her own function as shown by this part of a report on an eclipse in 1901, observed by a Dutch expedition to Sumatra:

'...I shall only mention the general mode of proceeding.

5 min before totality: Every man at his post and ready.

81 sec before totality: The word 'klaar' (ready) given by Mr Wackers, who states, on the ground glass of the 10-inch coronagraph, that the sun's crescent covers an arc of 90 degrees.

40 sec later: Timekeeper A calls out 'spiegels' (mirrors); the three assistants take the caps off the mirrors, their task being further to pay attention to the siderostats in order that no unexpected impediments might occur.

15 sec or 16 sec before totality: Call 'opgepast' (attention) either by Mr Muller, when the sun's crescent is seen to cover an arc of 45 degrees, or by the lookout, when the searchlight of the pigmy is screened off. The call 'opgepast' is to be repeated, in either case, by Timekeeper A. Beginning of totality, as observed by Mr Muller in the dark hut of the 40-foot lens. Call: 'los' (go). Timekeeper B begins the counting of seconds: nul, een, twee... Timekeeper A tries to estimate the interval between 'los' and 'nul' and notes the chronometer reading correspon-

ding to the count ten. Timekeeper B goes on counting up to 200 and is relieved by A: thus the 200th second is counted by both A and B. In case no sun is visible at the beginning of totality, 'nul' is called out at 12 h 19 m 51 s local civil time.
380 sec after the beginning of totality, or, in case the 2nd contact is lost, at 12 h 26 m 11 s; call 'sluiten' (shut) by Timekeeper B.
End of totality. Call 'over' by Mr Muller; Timekeeper A goes on counting till Mr Muller emerges from the dark hut and notes the chronometer reading.'

Sometimes such procedures were rehearsed up to 15 or 20 times to make sure all would go well on the day. Incidentally, the event just described was clouded out and yielded no useful results, although participants enjoyed the visual spectacle through light clouds.

The earliest record I have found of participation from southern Africa on a total eclipse expedition was when Mr JE de Villiers, an amateur astronomer from Cape Town, was a member of the British Astronomical Association's trip to the eclipse of August 1896 in Norway. De Villiers was a surveyor and provided the group with a meridian line for their observations, which in the end were also clouded out.

Responses to eclipses in some of these old records show the intense emotions they can provoke. A favourite description is that of Mabel Loomis Todd who was on board a yacht off the coast of Japan to view the same eclipse as that watched by De Villiers above:

'Grayer and grayer grew the day, narrower and narrower the crescent of shining sunlight. The sea faded to leaden nothingness . . . A penetrating chill fell across the land, as if a door had been opened into a long-closed vault. It was a moment of appalling suspense . . . the very air was portentous. The circling sea-gulls disappeared with strange cries . . .

Then an instantaneous darkness leaped upon the world. Unearthly night enveloped all. With an indescribable out-flashing at the same instant the corona burst forth in mysterious radiance. But dimly seen through thin cloud, it was nevertheless beautiful beyond description, a celestial flame from some unimaginable heaven . . . Simultaneously the whole northwestern sky, nearly to the zenith, was flooded with lurid and startlingly brilliant orange, across which drifted clouds slightly darker, like flecks of liquid flame, or huge ejecta from some vast volcanic Hades. The west and southwest gleamed in shining lemon yellow . . . The pale, broken circle of coronal light still glowered on with thrilling peacefulness, while nature held her breath for another stage in this majestic spectacle. Well might it have been a prelude to the shriveling and disappearance of the whole world – weird to horror, and beautiful to heart-break, heaven and hell in the same sky.'

Today, eclipses are eagerly awaited by observers worldwide, both scientific and casual. Prospects look favourable for viewing the southern African eclipses of June 2001 and December 2002, and we can look forward to seeing one of the greatest shows on Earth.

How eclipses occur

Solar eclipses

As the Moon orbits the Earth, it must at some time come between the Earth and the Sun. However, because the plane of its orbit is not in line with that of the Earth, it usually passes a little above or below the Sun, as seen from Earth. If that sounds complicated, look at the drawing below (figure 1).

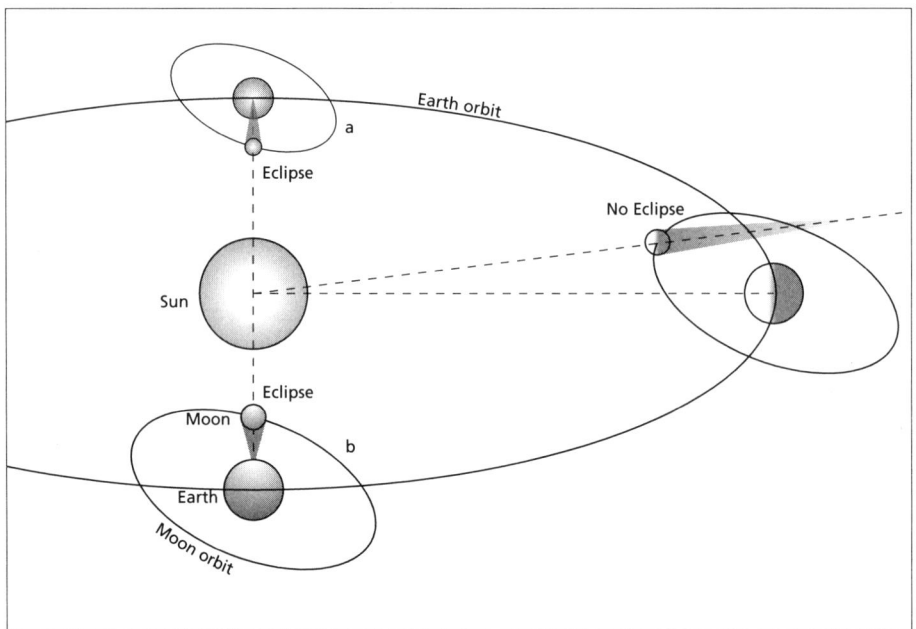

Figure 1: When the Moon and Sun are in line, one of them is usually a little above the other, except when at the positions shown as 'a' and 'b'.

Even when the Sun and Moon come to the positions shown as 'a' and 'b' in figure 1, the path of the Moon's shadow across the Earth may be almost entirely over the sea or over one of the polar regions that are inaccessible to observers. Because of this, the opportunity to view a solar eclipse is relatively rare for most people, although there are those who will travel worldwide to see the spectacle and to record it photographically or otherwise.

There are two further major complications: the relative sizes of the Sun, Earth and Moon and also their distances from each other. Because the Sun is about 100 times the size of the Earth, and the Moon only a quarter the size of the Earth, we get the peculiar shadow pattern shown in figure 2. In this instance, the Sun is totally hidden only from within the area marked 'a' (the umbra) and it is only partly obscured if seen from area 'b' (the penumbra).

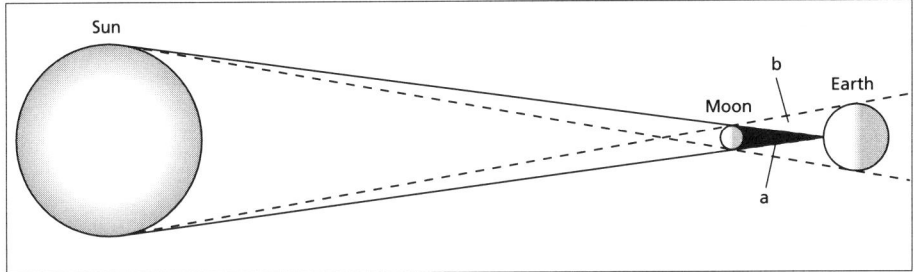

Figure 2: The Moon's shadow creates both an umbra (a) and a penumbra (b).

The fact that the umbra just stretches as far as the Earth is because of the coincidence that the Sun is both 400 times as far away and 400 times as big as the Moon. This size and distance relationship is what makes both Sun and Moon appear to be roughly the same size when we see them in the sky. In fact, the distances are not static because the orbits of the Earth around the Sun and of the Moon around the Earth are both ellipses and not true circles.

Therefore, if the Moon is at its furthest from Earth, the tip of its shadow will not quite reach us, and anyone directly in line with it will see a full ring of Sun all around the Moon; this is called an annular eclipse (see figure 3a). If the Moon is at its closest to the Earth, the size of the 'shadow spot' (caused by the umbra) on the surface of the Earth will be at a maximum (see figure 3b). Not only does this Earth-Moon distance determine the width of the track of totality (where the Sun is completely obscured), but it also contributes to the length of time the eclipse lasts. The other major factor deciding the length of totality is the speed of the shadow across the surface of the Earth. The changing distance between the Sun and the Earth also affects these issues, but the variation is so small in comparison with the huge overall distance, that its impact is minimal.

In Africa's two upcoming eclipses, the shadow spot for the June 2001 eclipse is much larger than for December 2002, and also has the longer period of totality.

When the Moon comes between us and the Sun, we are not able to see any detail on the Moon at all, as the side facing us is in darkness. What we will be observing is the Sun itself during the partial phases and the Sun's upper atmosphere and radiation effects during totality.

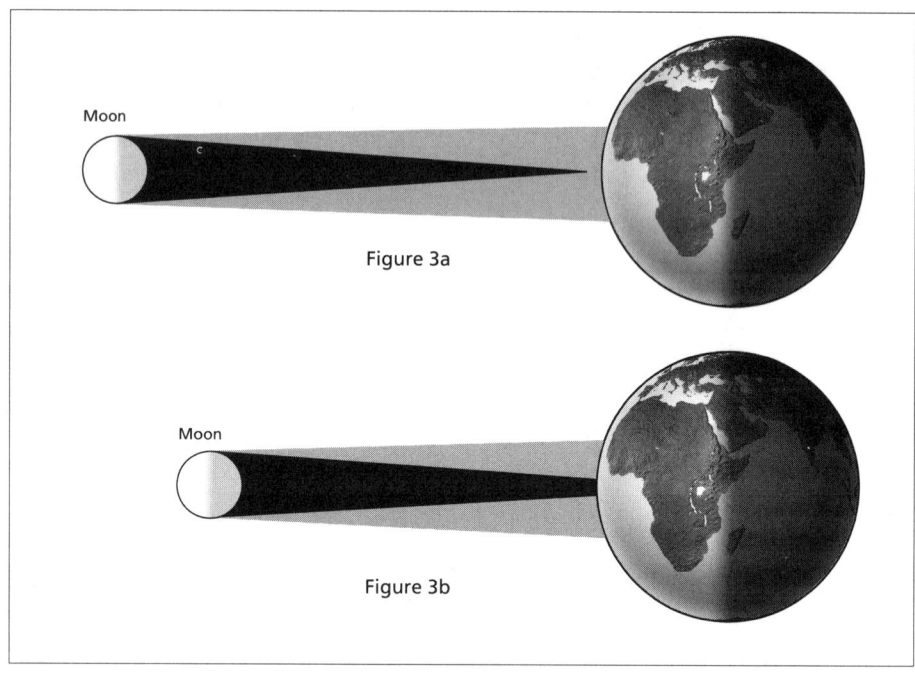

Figure 3a: *When the Moon is further from the Earth, the tip of its shadow does not reach us, and an annular eclipse occurs.*
Figure 3b: *When the Earth-Moon distance is smaller, a total eclipse of the Sun can be seen from within the shadow spot on the Earth.*

The Sun – our nearest star

The Sun is a great ball of burning gas and not a solid body. Consequently, explosions near the surface sometimes eject large volumes of hot gas for immense distances from its surface and these are best seen during total eclipses. These prominences can be clearly seen in several of the photographs in this book. Sometimes dark spots appear on the Sun and the ejected hot gases are associated with these (see plate 2). The spots appear dark against the rest of the solar surface, but in fact they have a temperature in the region of 3000 °C.

The sizes of the spots vary considerably, as also does the length of time for which they last. Some are quite puny and insignificant, but others are very large and display complicated shapes. The larger spots show a dark centre surrounded by a lighter shadow and we hope some will be visible during the partial phases of these eclipses. You will appreciate the size of the spots better if you remember that the diameter of the Sun is 100 times that of the Earth.

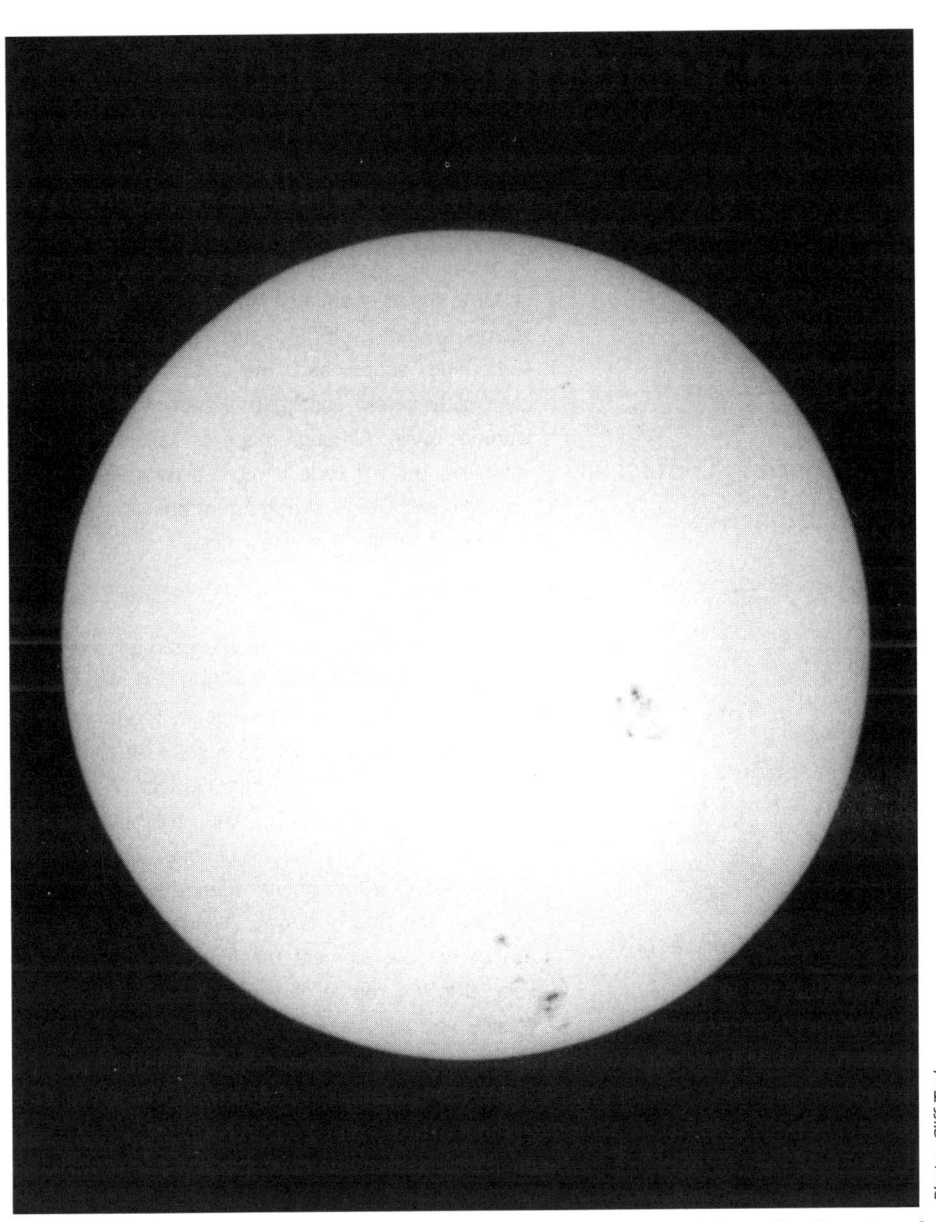

Plate 2: The Sun, showing large sunspots, some of which are up to 4 times the diameter of the Earth.

Sun

Moon

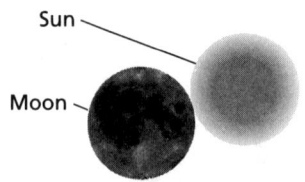

First Contact 4a

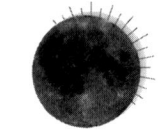

Second Contact 4b

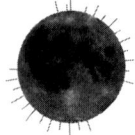

Totality

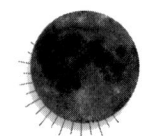

Third Contact 4c

Fourth Contact 4d

Figure 4

The Sun goes through a cycle of activity which builds up to a climax and then drops back to almost nothing every 11 years. This cycle is clearly shown by the number and location of the sunspots. At first they form at higher latitudes and are relatively infrequent and small. As time goes by, the number of sunspots increases and so does their size, and they tend to move closer to the solar equator. Finally, the whole sequence seems to get itself into such a tense state that something snaps and the spots vanish almost completely before coming back slowly in the higher latitudes again. Although this cycle takes 11 years to complete, the full cycle is more accurately 22 years, because each time the tension breaks there is a reversal of the Sun's magnetic field.

Order of events

The order of events for any eclipse is predicted by the times of 'contact' between the Sun and Moon as we see them from our chosen viewpoint. The moment the edge of the Moon touches the edge of the Sun (figure 4a) is called, appropriately, 'First Contact'. Then comes 'Second Contact' (figure 4b) when the Sun is blotted out totally for a short period of time. 'Third Contact' (figure 4c) is the moment when the Sun starts to reappear and totality is over. 'Fourth Contact' (figure 4d) occurs the moment the Moon finally clears the solar disc. Of course, if you are not in the area of totality you will see only First and Fourth Contacts. However you will still be able to watch the progress of the Moon across the Sun and with luck you should be able to see some of the sunspots mentioned earlier.

For instance, for the June 2001 eclipse as seen from South Africa, First Contact will be on the lower west side of the Sun, and the Moon will travel eastwards to leave the Sun at the lower east side

(Fourth Contact). The actual relative positions of Sun and Moon will vary as seen from different positions on the ground. As one travels further north, so the contacts also rise higher on the Sun until, in the line of totality, the Moon first touches the Sun close to centre west and progresses eastwards, finally leaving the Sun at centre east.

During the partial phases before Second and after Third Contacts, there is much to do. Most people will want to take some photographs of the Sun and also of fellow eclipse watchers. During this time, it is worth looking under any nearby trees where rays of sunlight will be shining through the gaps between leaves and creating numerous small eclipse images on the ground or convenient walls. The gaps act like dozens of pinhole cameras that each allow limited light to pass through, creating a multiple image effect. As for any pinhole camera, the image is inverted left-to-right and top-to-bottom – for example, a light ray from the topmost edge of the Sun enters the pinhole diagonally from the top and appears at the bottom of the projected image. This is easy to capture on a photograph, and makes an interesting souvenir (plate 3).

Photo: Ethleen Lastovica

Plate 3: Gaps between leaves act like dozens of pinhole cameras to create multiple images on the ground below of the semi-eclipsed sun.

Totality

When totality approaches, things get busy. If you are lucky enough to be on high ground, you will be able to see the Moon's shadow racing across the countryside towards you at about 100 kilometres per hour. As the shadow reaches you, the Sun will vanish from view and for a few seconds the Diamond Ring (a complete circle of light with just a small outburst of light at one point) will be seen, followed by what are known as Baily's Beads (pictured below) and then the full corona (the outer region of the Sun's atmosphere). There is no need for protective eyeglasses now that the Sun is completely hidden and you are in darkness. You may notice that birds have retired to their nests to roost and that it has become noticeably cooler. At this time too, some stars and planets will become visible and can be identified on a sky chart (see pages 34 & 51).

Photographs during totality are always a problem. Most of us get so excited that we don't think too clearly, so it is worthwhile planning your pictures beforehand. Short exposures will

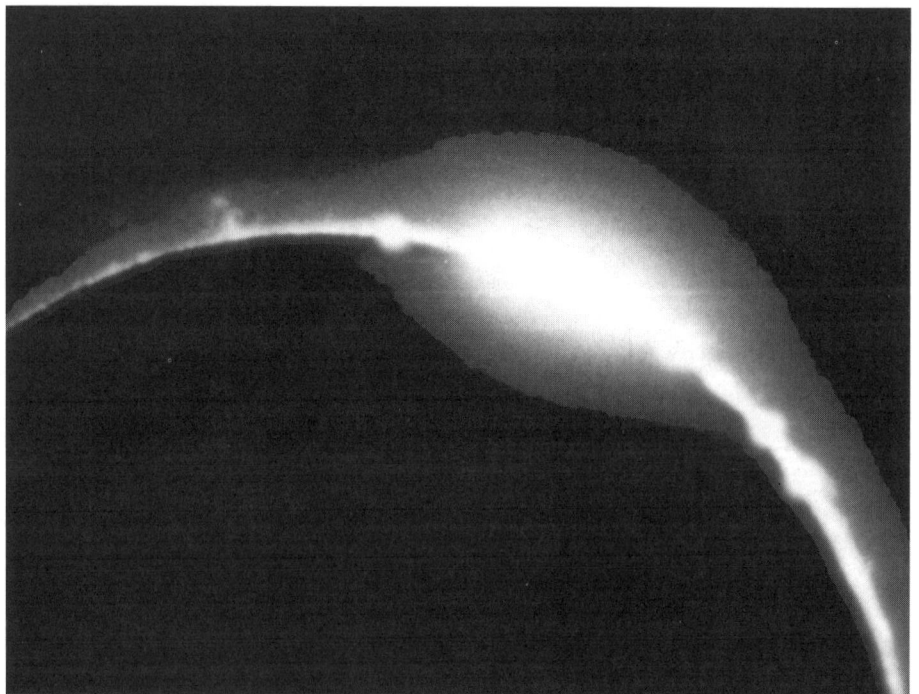

Photo courtesy of the Planetarium, SA Museum

Plate 4: Baily's Beads forming as the last of the Diamond Ring fades. A prominence shows to the upper left.

Plate 5: 'The corona burst forth in mysterious radiance . . . beautiful beyond description . . .'

show solar prominences, or burning material being ejected from the surface of the Sun. Longer exposures will bring out the corona further and further from the Sun, but too long an exposure will cause a blur unless your camera is specially mounted so that it tracks the Sun's movement across the sky (see page 56). About 3 minutes later, depending on your exact location, totality is over and you must be careful not to be caught looking at the Sun without safe eye protection when it first reappears.

The final phases

The final phases until Fourth Contact are almost an anticlimax, but everyone will be talking about their experiences during totality and the time will pass very quickly. One thing is certain. Once you have seen totality, you will want to see it again as soon as possible at the next accessible eclipse.

Lunar eclipses

When the Earth passes between the Sun and the Moon, a lunar eclipse can result. As the Earth is considerably bigger than the Moon, the shadow will also be bigger. The distance between the Earth and the Moon is relatively small, and the whole of the Moon can easily fit into the Earth's umbra, with no sunlight falling onto the Moon's surface (see figure 5a).

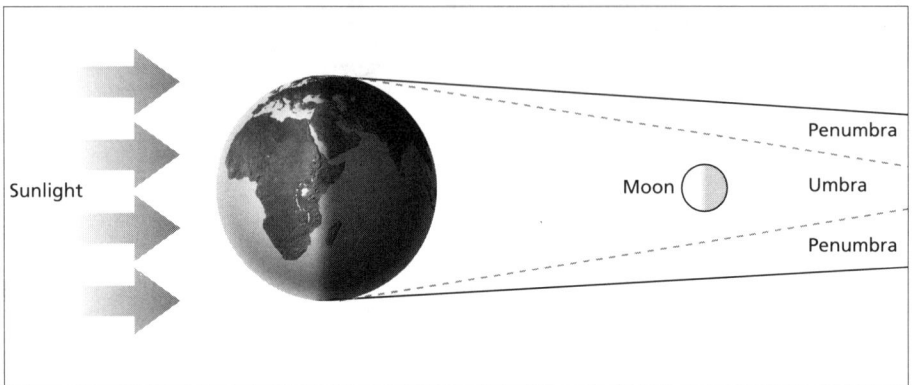

Figure 5a: The Earth's shadow eclipses the Moon completely.

A lunar eclipse takes place when the shadow of the Earth falls upon the surface of the Moon. Any shadow that is cast by an extended light source, such as the sun, (as opposed to that cast by a point light source) has two distinct areas – and that of the Earth is no exception. The central shadow, or 'umbra', is the darkest and will be easily seen on the Moon, but this is surrounded by a much lighter partial shadow known as the 'penumbra' which is often hardly noticeable at all when the Moon passes through it. However, the penumbra does reduce the amount of light reaching the Moon, and this in turn reduces the reflected light.

When sunlight passes through our atmosphere as it travels past the Earth, it is scattered by dust particles, and both scattered and refracted (bent) by the atmosphere itself. The effect of this is that we see much more of the red part of the spectrum of the light, and this is what gives us red sunsets. The same effect enables us to see the Moon even when it is fully eclipsed by the Earth's shadow and it also produces the rusty red colour so often seen on the eclipsed Moon (see figure 5b).

Because the light refracted onto the Moon is not direct sunlight, it results in consider-ably softer shadows than if it were direct. The edge of the Earth's advancing shadow has a much softer edge than we see in solar eclipses. Regular lunar eclipse watchers like to

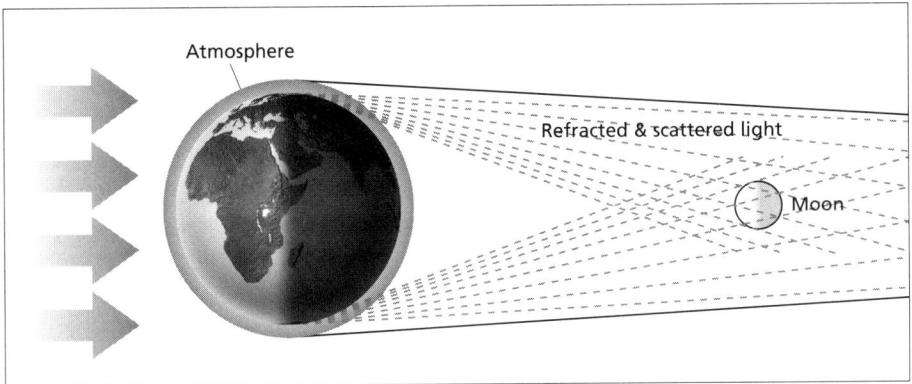

Figure 5b: Sunlight refracted (bent) by the Earth's atmosphere illuminates the Moon even though it is totally within the shadow cast by the Earth.

record the exact time at which the shadow reaches various features on the Moon, but the soft edge of the shadow makes this quite difficult and it requires much practice.

It is quite possible for the Moon to move through the penumbra of the shadow of the Earth without touching the umbra at all. This is still an eclipse, but there is no clear division between light and dark areas and, as a result, interest in such eclipses is not high.

Although there is normally little colour in the Moon and black and white photography is adequate, an eclipse is the exception to this rule because of the refracted sunlight and orange-red tinting, which is often very beautiful and impressive. Determining the best exposure times is difficult because of varying depths of the eclipses and one should be prepared to use lots of film – this is discussed in more detail in the chapter on Photographing eclipses.

Eclipse observing

Solar eclipses

This chapter may appear to try to put the reader off eclipse viewing for ever. This is not the intention, but it must be stressed that the Sun produces very intense radiation throughout the spectrum and can be extremely dangerous for the unwary. No apologies are made for repeated warnings.

The golden rule for looking at the Sun is never to take any chances with eye protection that you do not *know* to be safe.

Some filters that are NOT SAFE:
(i) Smoked glass
(ii) Exposed photographic film
(iii) Dark photographic or darkroom filters
(iv) Ordinary sunglasses (including polaroids)
(v) So-called 'sun filters' sold with small telescopes
(vi) Compact discs

Naked eye viewing and precautions

Many people do not realize that the Sun is a star and it is the only one near enough to us to be seen in detail. Because of its closeness, it is very bright indeed and looking directly at it – even for only an instant – can cause eye damage which cannot be corrected and could result in blindness. NEVER take chances. Your eyes are too precious to put at risk.

Small telescopes are often sold with a dark filter that is labelled 'Sun Filter', or something similar. These filters fit into or onto the eyepiece of the telescope and are alleged to be safe for direct viewing of the Sun. They might appear to be adequate, but they are NOT safe, and are best destroyed at once.

There can be significant heat build-up at the eyepiece of a telescope pointed at the Sun and this can cause the filter to crack and let the full force of the Sun's energy through to your eye. The author has twice had personal experience of this problem, though luckily nobody was injured on either of these occasions as the observer had turned away from his telescope just before the filters cracked. Others have not always been so lucky.

It *is* possible to use totally exposed black and white (not colour) photographic film which has been completely developed or even over-developed, but then two or even three thicknesses should be used at once. Ensuring it doesn't become accidentally scratched after development can be a nuisance. Because of this fragility, other methods are preferred.

Welders' glass of grade 14 or darker (i.e. higher numbers) is much more convenient. This is not easily available from welding suppliers who normally only stock around numbers 8 to 10 – but can be obtained if ordered well in advance. (The author has a small supply of grade 14 welders' glass for sale – see contact details on page 64.) This glass gives a green image of the Sun. Once acquired, welders' glass is best fixed into a piece of masonite or something similar to make a window in the surrounding protection. A welders' helmet would be ideal, but be sure to substitute your new grade 14 glass for its usual glass.

Various types of commercially produced protective glasses and filters are available, but at the time of writing, it is not clear which ones will become available in southern Africa. There are two main types of filters which show either a blue or an orange image of the Sun. Those from Europe should be marked as safe for direct solar viewing and carry a 'CE' mark, which indicates that the filter has been passed under the provisions of Council Directive 89/686/EEC of the European Union or similar regulations in the United Kingdom.

The South African Astronomical Observatory is arranging to have a large number of filters manufactured locally, and hopes to be able to distribute them in newspapers as well as to schools. The author has seen a sample of these solar screen filters, which show an orange image, and they are excellent. At the time of writing, purchasing details are not known. Further information can be obtained on the Observatory's web page (see page 64).

Mylar, obtainable from packaging companies, offers excellent protection. As it is available in differing densities when used for packaging, it is a good idea to use it double thickness which also avoids any possible trouble from tiny pinholes. It is a thin plastic film coated with an aluminium surface, which reflects a high percentage of light. One side often seems to be more highly polished than the other. Put the high polish towards the light source (the Sun) so that the maximum amount of light is reflected and not allowed to penetrate through the film. Something of the order of 99.99% or more should be stopped (i.e. reflected). Mylar looks somewhat like kitchen foil, but cannot be torn as easily. Make sure you have the genuine product, which can also serve as a filter for your camera as described in the chapter, 'Photographing eclipses'.

Binocular and small telescope viewing

Never look directly at the Sun through any telescope or binoculars unless you know exactly what you are doing and have experience using the correct type of filters in the correct way. In most cases, this means having an experienced astronomer to guide you.

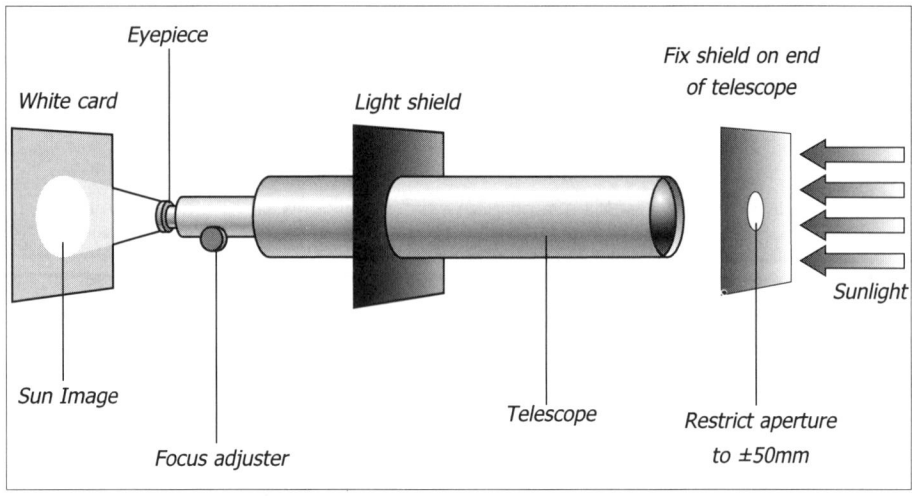

Figure 6: Projecting the Sun's image through a telescope on to a screen.

Photo: Ilona Dicker

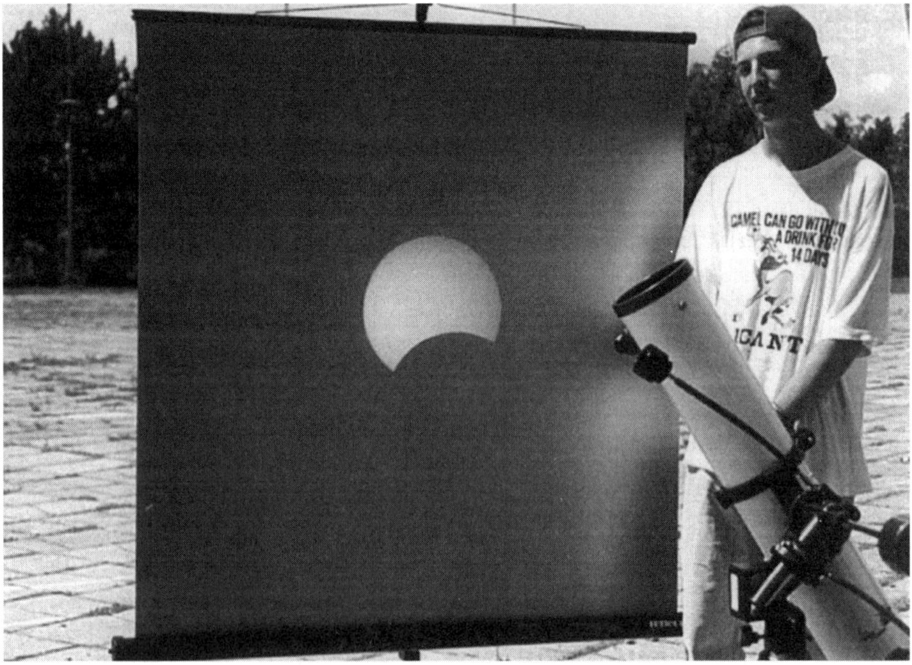

Plate 6: The solar eclipse of 1999, is projected through a telescope on to a screen.

The safest way is to view the Sun by projection, using a telescope or one barrel of a pair of binoculars, and for which a 50 mm aperture is ample. A large telescope is not needed, and can be a problem as it has to be stopped down to about 50 mm (maximum 60 mm) to avoid damaging the eyepiece from heat build-up. Once the binoculars or telescope are mounted firmly on a tripod, simply place a screen beyond the eyepiece on which to focus an image of the Sun. (See figure 6 for a diagrammatic explanation of the setup, and plate 6).

Amateur astronomers will be familiar with special filters that cover the whole aperture of telescopes. These may take the form of Mylar (which gives a blue image) or Iconel (which gives an orange image), among others, some of which only filter out certain light wavelengths. Used correctly, in accordance with manufacturers' specifications, they are excellent – but expensive for anyone who is likely to use them only once. They are, therefore, outside the scope of this book which is for the casual onlooker rather than committed amateur astronomers.

Viewing in a box

A safe and easy way to view a solar eclipse is with the use of a box – any medium-sized cardboard box will do that has the lid or top removed. Punch a small hole (1–2 millimetres in diameter) in one of the sides of the box. For viewing the eclipse, hold the box in such a way that the Sun shines through the pin-hole and casts an image on the opposite inside wall of the box. This projected image of the Sun can be viewed through the open panel of the box. The image will be inverted top-to-bottom and left-to-right because of the pinhole-camera effect, and will clearly show the phases of partiality until the eclipse becomes total, when no further direct light reaches the Earth.

A similar arrangement can be set up using a telescope. Once again, a hole must be made in one side of the box, this time large enough to contain the eyepiece of the telescope. In this case, a magnified image of the Sun will be projected on to the far side of the box. Not only is this a perfectly safe method of viewing the Sun, but it also enables a group of people to enjoy the show. Plate 7 on page 24 shows how a telescope and box can be set up to achieve this effect.

Objective filtering for direct viewing

If you have been able to obtain Mylar and are certain that it is the genuine article, you can mount two or three layers of it over the object lenses (larger lenses) of binoculars (NOT over the eyepieces), but make sure the Mylar layers are well fixed in place and cannot be blown or knocked off. You can then view the eclipse directly, but only do this if you are certain of what you are doing. If in any doubt, start with more layers of Mylar – they can always be reduced one at a time if your view is totally blocked out. Don't have the image too bright;

Photo: Audrey Joubert

Plate 7: An image of the Sun is projected through a telescope into a box for safe viewing.

remember, you will be looking at it almost constantly for several hours. Similarly, layers of Mylar can be fixed over the sunward end of a telescope without restricting its aperture to the 50 mm mentioned above, but again, make absolutely certain that it cannot come off accidentally, as the results could be tragic.

You should see sunspots on the Sun's surface during the partial phases, but you will not be able to see any surface features during totality. You will not have much time to remove the filtering material and look at totality before the Sun reappears – possibly with disastrous results – so do not attempt such a rushed adjustment. If possible, have a second camera or telescope set up for viewing just totality – without filtering material, but protected from the long wait in the sun by a lens cap.

Partial phases

Watching the Moon slowly cover the Sun from First Contact is fascinating and, with any luck, there will be some big sunspot groups that are easily visible to the naked eye. Those using binoculars or small telescopes for projection of an image onto a screen will almost certainly see more sunspots. This will also allow other people who are not lucky enough to have their own optical aid to see sunspots too.

Totality

This is the only time during the whole duration of the eclipse when you do not need any filters at all, either for viewing or photographing. When the Moon has completely covered the Sun, you see the real magic and beauty of the solar eclipse. This is what it is all about. But remember that you have limited time, and it is essential that you check beforehand with the predictions so that you know how long totality lasts (see 'totality duration' in the detailed tables, starting on page 28). The author carries with him an accurate darkroom clock pre-set to about 10 seconds less than the predicted duration of totality and starts it as soon as totality begins. It then rings to give him 10 seconds warning of the return of direct sunlight. This works well, but it is easy to forget to start the clock immediately because you are so absorbed with the spectacle – especially the first time you see totality – and you would be wise to allow a longer period than 10 seconds for safety.

During totality, the corona becomes visible. It is not possible to predict what it will look like as all eclipses differ, but it is always spectacular. The size and complexity of the corona is determined by how active the Sun's surface is at the time. The amount of solar activity varies considerably, almost on a daily basis, but in the longer term it follows a regular cycle. This cycle takes a period of 11 years and we are lucky that the Sun is currently nearing the maximum of its current cycle. We can therefore expect magnificent views of the corona, but predictions are difficult and anything could happen on the days of the eclipses.

Getting comfortable

Some time before the eclipse, it is worthwhile spending a little time checking on the antici-pated position of the Sun at eclipse time of day, so you know the elevation and direction in which you will be looking. You can then choose whether to take a groundsheet and cushions so you can lie on the ground to watch high in the sky, or a reclining chair if it is at a lower angle. This applies equally to observing through mounted binoculars or small telescopes, and sometimes a 90-degree angle device known as a 'star diagonal' can direct telescope projec-tions to more convenient positions. A comfortable folding chair is always useful as you could be standing around for several hours.

Remember too that in December 2002 it could be very hot and a beach umbrella for shade would make conditions more bearable. Suntan lotion will almost certainly be needed on your face, which will be turned towards the Sun for long periods. Don't forget a supply of liquid refreshment, especially if you have children with you. You don't want to be caught queuing for drinks as totality arrives.

June 2001 shouldn't be too hot, but may produce cold winds, and wind protection will improve your comfort, as also will a flask of hot coffee.

Lunar eclipses

Anyone anywhere on the side of the Earth facing the Moon is able to see the Moon – and watch its eclipses. With a lunar eclipse, there is no problem about a 'narrow band of totality', and no need to travel to special locations. Factors affecting viewing, based on the observer's geographic position, will be the position of the Moon in the sky and prevailing weather conditions. No eye protection is needed to look at or photograph a lunar eclipse. After all, even the fully lit Full Moon does not harm our eyes, so the Moon in shadow is even less harmful.

When the Moon enters the penumbra of the Earth's shadow, there is no noticeable difference in its brightness, (figure 7a) but as it approaches the umbra, observers may notice that it seems less bright than previously. On reaching the umbra, a definite shadow becomes visible on the Moon, although this shadow does not have a hard, sharp edge – it is always a little fuzzy. The Moon may pass only through the penumbra which provides very little to see at all, or it may partly immerse itself in the umbra, giving a partial eclipse (figure 7c).

The third possibility, when the Moon goes fully into the umbra, is a total lunar eclipse (figure 7b). To observe total or partial eclipses, it is wise to arm oneself with a large-scale map of the Moon (included in most star atlases) with all the craters, mountains and other features clearly marked for identification. The advancing shadow can then be timed as it reaches the various features – and again when it finally uncovers those same features.

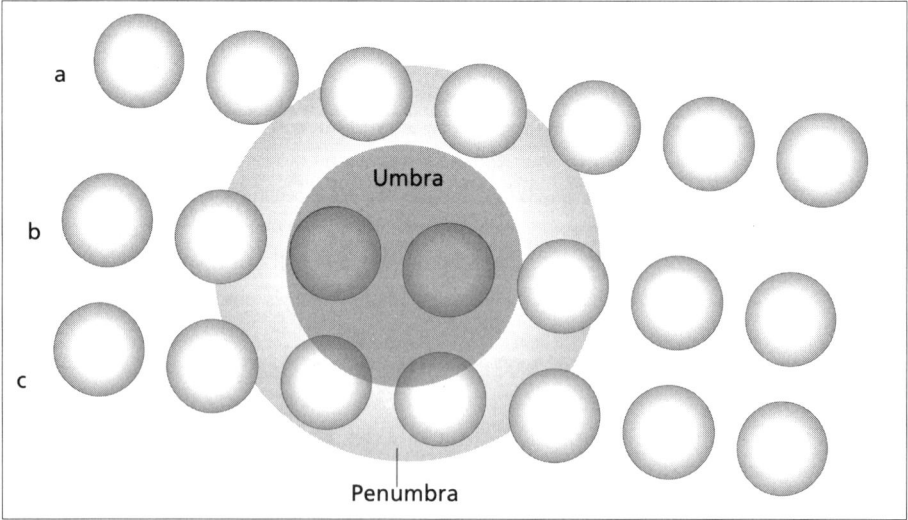

Figure 7: Diagram of the different paths the Moon can take through the Earth's shadow, resulting in three different types of eclipse: (a) Penumbral, (b) Total, (c) Partial.

Eclipse of 21 June 2001

Overview

In June 2001 a total eclipse of the Sun will pass over southern Africa, creating a narrow band of totality that will traverse Angola, Zambia, Zimbabwe, Mozambique and Madagascar, and will be visible as a partial eclipse from a dozen more countries. The centreline of this eclipse will traverse a part of Africa that has a dry winter season, when the amount of sunshine during daylight hours will probably be in the region of 70 to 80%. Detailed below are viewing possibilities from the principal countries, as well as probable weather conditions, effects of pollution, accessibility and the extent of tourist facilities.

> Table 1 (page 28) details the path of totality for the eclipse of 21 June 2001, its duration and the predicted times of the various contacts as described in the previous chapter. Partial eclipse times are given in Tables 2a–d, with countries arranged in alphabetical order. Maps 3–6 on pages 36–43 show the geographical path of totality.

Angola

Weather conditions on the Angolan plateau seem to be among the best anywhere on the path of totality. However, it would seem unlikely that many people will travel to Angola to view this eclipse, mainly because of the civil war that has been continuing in that country for many years. All weather predictions in this area are based on satellite observations alone, as there are virtually no ground-based stations noting weather conditions.

It is unfortunate that political instability makes Angola an unlikely viewing venue, as not only does it present the best weather prospects, but totality lasts longer on this western side of Africa, where it can extend to over 4 minutes 30 seconds.

During winter, the conditions here are generally clear, with cloud cover being less of a problem than smoky haze, which is caused by biomass burning. This haze is more concentrated north of the path of totality but could still cause some difficulty in photographing or viewing the outer parts of the corona during totality.

Zambia

Winter months in this area bring virtually no rain, although an influx of cold air along the Limpopo River can bring occasional showers. The haze described above for Angola, caused by biomass burning, can also be a problem in Zambia, but to a lesser degree.

Table 1: Local details of the total eclipse of 21 June 2001

NB: All times are quoted in Universal time (previously GMT) and should be corrected to local time by adding the following: Angola: 1 hour; Zambia, South Africa & Zimbabwe: 2 hours; Mozambique: 3 hours

Site	1st Contact h m s	Alti-tude	2nd Contact h m s	3rd Contact h m s	4th Contact h m s	Alti-tude	Totality duration
ANGOLA							
Chibango	11:23:10	51	12:57:34	12:59:52	14:20:45	23	2m 18s
Conda	10:59:15	55	12:37:45	12:42:17	14:09:53	33	4m 32s
Ngunza	10:57:35	55	12:36:15	12:40:50	14:08:55	34	4m 35s
Porto Amboim	10:57:25	56	12:36:22	12:40:28	14:08:53	34	4m 06s
ZAMBIA							
Chavuma	11:25:37	51	12:58:34	13:02:33	14:21:53	33	3m 59s
Chisamba	11:41:58	45	13:09:31	13:13:02	14:27:15	16	3m 32s
Kabwe	11:42:19	46	13:10:37	13:12:34	14:27:30	16	1m 57s
Kafue	11:41:10	45	13:09:52	13:11:17	14:26:43	16	1m 25s
Kambanga	11:26:43	51	12:59:20	13:03:16	14:22:15	22	3m 56s
Lusaka	11:41:34	45	13:09:19	13:12:33	14:27:00	16	3m 14s
Mukinge Hill	11:35:14	49	13:06:11	13:08:04	14:25:22	19	1m 53s
Mumbwa	11:38:15	46	13:07:13	13:10:31	14:26:01	17	3m 18s
Mushima	11:32:41	49	13:03:30	13:07:04	14:24:17	20	3m 35s
Rufunsa	11:45:26	44	13:12:06	13:14:48	14:28:18	15	2m 42s
ZIMBABWE							
Bradley Inst	11:49:25	41	13:14:51	13:16:08	14:28:51	12	1m 16s
Chipuriro	11:47:38	42	13:13:29	13:15:28	14:28:27	13	1m 59s
Chirundu	11:43:04	44	13:10:48	13:12:43	14:27:16	15	1m 55s
Makaha	11:52:14	40	13:15:54	13:18:23	14:29:34	11	2m 29s
Rusambo	11:51:28	41	13:15:11	13:18:30	14:29:36	12	3m 19s
MOZAMBIQUE							
Changara	11:53:53	13	13:16:37	13:19:50	14:30:10	10	3m 13s
Chinde	12:00:24	35	13:20:10	13:23:01	14:31:10	07	2m 52s
Queliman	12:01:42	35	13:21:44	13:23:18	14:31:44	07	1m 33s
MADAGASCAR							
Ambahikily	12:12:39	26	13:26:01	13:28:40	-	-	2m 39s
Andranopasy	12:12:55	26	13:26:22	13:28:48	-	-	2m 26s
Farafangana	12:18:07	22	13:28:58	13:30:13	-	-	1m 15s
Morombe	12:12:02	26	13:25:48	13:28:10	14:32:02	00	2m 22s

Table 2a: Local details of the partial eclipse of 21 June 2001

NB: All times are quoted in Universal time (previously GMT) and should be corrected to local time by adding the following: Angola: 1 hour; Zambia, South Africa & Zimbabwe: 2 hours; Mozambique: 3 hours

Site	1st Contact h m s	Alti-tude	Max eclipse h m s	Alti-tude	Max % obscured	4th Contact h m s	Alti-tude
ANGOLA (see also table 1 for totality)							
Benguela	10:55:54	54	12:36:49	48	98,7	14:07:33	33
Huambo	11:03:34	54	12:43:30	45	99,2	14:11:49	41
Lobito	10:56:26	54	12:37:21	48	99,4	14:07:57	33
Luanda	10:56:35	58	12:37:37	51	96,1	14:08:23	36
BOTSWANA							
Gabarone	11:34:17	38	13:01:00	28	69,7	14:16:23	15
CONGO							
Brazzaville	11:06:24	62	12:44:28	52	78,5	14:11:25	36
Pointe-Noire	10:54:52	61	12:34:44	56	82,2	14:05:30	40
DEMOCRATIC REPUBLIC OF CONGO							
Kinshasa	11:06:25	62	12:44:31	52	78,7	14:11:27	36
Kolwezi	11:35:08	51	13:07:10	37	92,2	14:25:30	21
Lubumbashi	11:40:39	49	13:10:44	34	93,1	14:27:16	18
GABON							
Libreville	10:53:28	66	12:29:01	62	63,3	13:57:57	46
GHANA							
Accra	10:38:42	63	12:02:57	72	41,5	13:28:25	63
IVORY COAST							
Abidjan	10:31:00	59	11:52:14	71	39,2	13:17:09	67

Access to the path of totality is quite easy, with Lusaka being well within the path of the eclipse, and enjoying over 3 minutes of totality. From Lusaka, the Great West Road (M9) is within the southern half of totality for about 300 kilometres. It is the main access road to the Kafue National Park, and a number of safari trips will be on offer for eclipse watching. Further north, the roads in the Park may be soft and a vehicle with 4-wheel drive could be needed to reach the centreline of the eclipse. The Great East Road crosses the centreline about 60 km from Lusaka and has the reputation of being one of the best roads in the country. Two other roads leading north also cross the centreline within 80 km of the capital.

Table 2b: Local details of the partial eclipse of 21 June 2001

NB: All times are quoted in Universal time (previously GMT) and should be corrected to local time by adding the following: Angola: 1 hour; Zambia, South Africa & Zimbabwe: 2 hours; Mozambique: 3 hours

Site	1st Contact h m s	Alti-tude	Max Eclipse h m s	Alti-tude	Max % obscured	4th Contact h m s	Alti-tude
KENYA							
Mombasa	12:18:53	39	13:28:41	24	47,6	14:29:24	11
Nairobi	12:15:22	44	13:25:23	29	42,3	14:26:20	15
LESOTHO							
Maseru	11:38:39	33	13:00:06	23	58,1	14:11:46	12
MALAWI							
Blantyre	11:58:34	38	13:21:08	24	96,8	14:31:37	9
Lilongwe	11:56:25	41	13:20:06	26	92,5	14:31:23	11
MOZAMBIQUE (See also table 1 for totality)							
Beira	11:56:34	36	13:18:59	22	96,4	14:29:35	8
Maputo	11:50:03	33	13:11:31	21	73,7	14:22:08	8
Tete	11:55:00	40	13:19:02	25	99,8	14:30:40	10
MADAGASCAR (See also table 1 for totality)							
Antsirabe	12:19:02	24	13:30:51	10	92,3	-	-
Fianarantsoa	12:18:00	23	13:29:59	09	97,6	-	-
Toliara	12:11:37	25	13:26:11	12	97,2	-	-
NAMIBIA							
Windhoek	11:08:42	44	12:42:11	37	68,9	14:05:32	24
NIGERIA							
Lagos	10:48:49	68	12:13:30	72	40,2	13:36:50	59

Over a 35-day test period during which prevailing weather conditions were recorded, there was cloud cover on only 7 days, when early morning cloud dispersed and then damp air brought it back again around noon. It was found that the quicker the early cloud evaporated, the less cloud formed later in the day. The remaining 28 of the 35 days of the test showed no cloud that would interfere with eclipse viewing. However, even if there is some cloud at eclipse time, all is not lost. Some very pleasing photographs have been taken through light cloud, which can add to the effect (see the photograph of the Diamond Ring on page 33).

Table 2c: Local details of the partial eclipse of 21 June 2001

NB: All times are quoted in Universal time (previously GMT) and should be corrected to local time by adding the following: Angola: 1 hour; Zambia, South Africa & Zimbabwe: 2 hours; Mozambique: 3 hours

Site	1st Contact h m s	Alti-tude	Max Eclipse h m s	Alti-tude	Max % obscured	4th Contact h m s	Alti-tude
SOUTH AFRICA							
Alexandria	11:36:48	30	12:54:03	22	46,7	14:03:11	12
Benoni	11:40:40	35	13:04:31	24	67,8	14:17:27	12
Bloemfontein	11:35:21	34	12:57:45	25	57,5	14:10:21	13
Boksburg	11:40:09	36	13:04:13	25	67,9	14:17:21	12
Cape Town	11:17:46	32	12:36:52	27	41	13:49:29	18
Carletonville	11:38:04	36	13:02:30	25	66,5	14:16:04	13
Daveyton	11:40:35	36	13:04:35	25	68,3	14:17:38	12
Durban	11:46:12	31	13:05:32	20	60,4	14:15:08	9
East London	11:39:58	30	12:57:27	21	49,5	14:06:26	10
Evaton	11:39:22	36	13:03:22	25	66,6	14:16:32	12
Germiston	11:39:59	36	13:04:05	25	67,8	14:17:15	12
Johannesburg	11:39:35	36	13:03:46	25	67,5	14:17:01	12
Kempton Park	11:40:09	36	13:04:18	25	68,2	14:14:29	12
Kimberley	11:31:59	35	12:55:32	26	57,5	14:09:15	15
Klerksdorp	11:36:21	36	13:00:41	26	64,1	14:14:22	13
Mamelodi	11:40:18	36	13:04:42	25	69,3	14:18:01	12
Mdantsana	11:39:30	30	12:57:08	21	49,5	14:06:16	11
Natalspruit	11:39:57	36	13:03:59	25	67,5	14:17:06	12
Pietermaritzburg	11:44:48	31	13:04:44	21	60,5	14:14:51	9
Port Elizabeth	11:35:15	30	12:52:22	22	45,4	14:01:33	12
Pretoria	11:40:00	36	13:04:29	25	69,1	14:17:53	12
Soweto	11:36:21	36	13:03:36	25	67,5	14:16:55	12
Thohoyandou	11:45:42	37	13:10:28	24	80,3	14:23:28	11
Umtata	11:41:41	30	13:00:22	21	53,8	14:09:58	10
Vereeniging	11:39:29	35	13:03:22	25	66,4	14:16:26	12

Zimbabwe

The eclipse path crosses northern Zimbabwe, the centreline remaining within Zimbabwe for about 370 km (except for one relatively small break near the west of the country), before moving into Mozambique. The length of totality on the centreline remains at over 3 minutes, but the further east one goes, the more chance there is of cloud interference. Satellite obser-vations spanning more than 10 years show that cloud cover could be in excess of 30% at the eastern end of the track in Zimbabwe. However, there is little chance of interference from biomass burning (as there might be in Angola and Zambia), and much of the track lies along the Zambezi escarpment at heights up to 1 500 m, above the valley cloud.

The tourist industry in Zimbabwe is better developed than in the other countries traversed by the eclipse, and there is accommodation in safari camps and elsewhere all along the northern border. The Harare Centre of the Astronomical Society of Southern Africa set up an eclipse sub-committee at an early stage and is able to supply help and advice on availability of accommodation and accessibility. (See 'Contacts & Addresses' on page 64).

Most roads in the area run in a north-south direction, which makes it difficult to make a last-minute dash to the east or west along the track of the eclipse if the weather is unfavourable – not that such relatively short distances would make a difference, anyway.

Table 2d: Local details of the partial eclipse of 21 June 2001

NB: All times are quoted in Universal time (previously GMT) and should be corrected to local time by adding the following: Angola: 1 hour; Zambia, South Africa & Zimbabwe: 2 hours; Mozambique: 3 hours

Site	1st Contact h m s	Alti-tude	Max Eclipse h m s	Alti-tude	Max % obscured	4th Contact h m s	Alti-tude
SWAZILAND							
Mbabane	11:46:46	34	13:08:58	22	70,9	14:20:19	9
TANZANIA							
Dar es Salaam	12:15:01	39	13:28:16	23	57,9	14:31:33	9
Tabora	12:00:26	47	13:20:22	32	61,3	14:28:49	17
Zanzibar	12:15:26	39	13:28:12	24	55,8	14:31:07	10
ZAMBIA (See also table 1 for totality)							
Chingola	11:41:25	48	13:11:13	33	95,5	14:27:30	18
Kitwe	11:42:18	47	13:11:45	32	96,1	14:27:45	17
Luanshya	11:42:41	47	13:11:58	32	96,9	14:27:50	17
Mufulira	11:42:28	47	13:11:52	33	95,1	14:27:48	17
Ndola	11:43:25	47	13:12:26	32	96,0	14:28:04	17
ZIMBABWE (See also table 1 for totality)							
Bindura	11:49:02	41	13:15:11	27	99,5	14:28:37	12
Bulawayo	11:41:23	41	13:09:08	28	87,1	14:24:17	14
Harare	11:48:11	41	13:14:30	27	97,8	14:28:08	12

Mozambique

This is likely to be the most cloudy part of the eclipse path, with sunshine reduced to 6 to 8 hours per day, depending on whether one is near the coast or further inland. The coastal areas have the least cloud but access by road is more difficult further inland to the north. The highway from Harare in Zimbabwe, via Tete in Mozambique, to Blantyre in Malawi is kept in good condition and crosses the centreline of totality only a few kilometres from the Zimbabwe/Mozambique border. In spite of the possibility of cloud interference, it is likely that this will be the most popular place for eclipse watching, especially for those travelling from Malawi and other places further north.

There is frequent overnight fog on the coast at Quelimane, and cloud may develop in the vicinity in the early afternoon when the eclipse is due. Thus the general attraction of the less cloudy coastal region is offset by the likelihood of fog and its residue, as well as more difficult road access.

Photo: Cliff Turk

Plate 8: Even in overcast weather it is possible to capture a good picture of the Diamond Ring.

Madagascar

Although the eclipse reaches Madagascar later in the day and is cut short by sunset, the weather conditions for the west side of the island are bettered only by those in Angola and western Zambia.

The trade winds from across the Indian Ocean gather much moisture, but this is precipitated along the eastern side of the island when the air is forced to rise over the mountains.

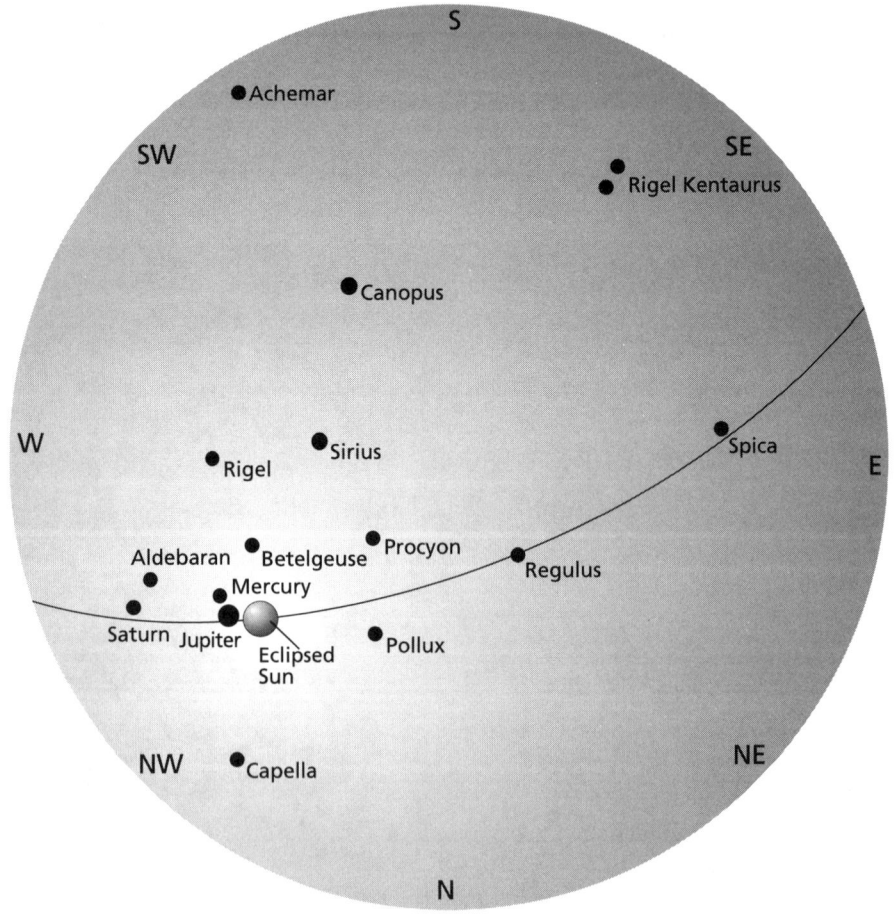

Figure 8: Positions of the planets and the stars during the eclipse of 21 June 2001.

As it descends on the western side, the air is generally dry and the weather clear and sunny.

Transport poses some difficulty as the centreline of totality crosses the coastal zone some 210 km north of the major town of Toliara. The road journey from Toliara northwards to Morombe is 306 km and can take up to two days at speeds of no more than 30 km/h. An alternative route is for 4-wheel drive vehicles only, and a guide is needed. Fortunately, Morombe has an airport and it seems a better proposition to arrange to fly in.

The centreline is still a further 37 km northwards at the village of Ambahikily on the banks of the Mangoky River but the view to the west is restricted, so the final moments of the eclipse would not be seen. Morobe offers better westerly viewing from its beach, but the length of totality drops to well under 2,5 minutes.

Sunshine at Toliara during June averages around 9 hours per day which is 85% of the total possible. June rainfall in Morobe in recent years has varied between 9 mm and zero. South-east Madagascar is likely to have more of a problem with the weather, as it can expect only 66% sunshine and could have broken cloud for well over 50% of daylight hours. Anyone travelling this far would be well advised to take a little more time and visit the south-western corner of the island.

Other countries in southern Africa

It is not necessary to cover the weather prospects for all the countries where there will only be a partial eclipse, as most observers in those places will be local residents who already know what to expect of the weather.

Predicted times of start and finish of the partial eclipse for many towns and cities throughout southern Africa are given in Tables 2a to 2d, together with the time of maximum eclipse at each place and the percentage of the Sun that will then be obscured.

Stars and planets visible during totality

During totality, there are a number of bright stars and planets that should become visible. Jupiter will be the most prominent as it is only 5° west of the Sun. It should be quite easy to identify, but any doubt can be cleared up by remembering that the Sun's (or Moon's) diameter represents about half a degree. Alternatively, your fist held at arm's length is about 10° across. Saturn will be over 22° from the Sun, also to the west and less than half as bright as Jupiter. At arm's length, your widespread hand covers about 20° from thumb to little finger. The bright star Sirius will be well above and to the right of the Sun.

Unfortunately, the brilliant Venus will set shortly before totality, except in Angola and the most easterly parts of Zambia.

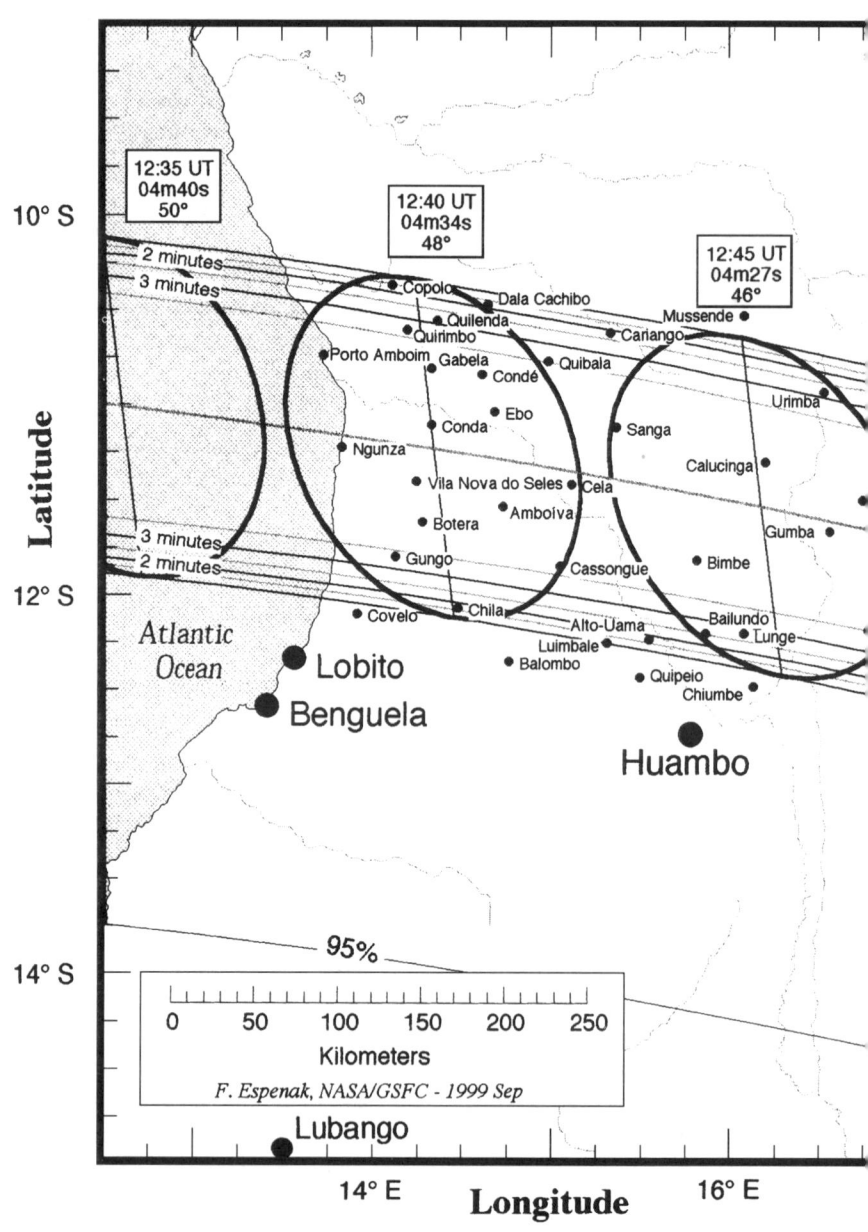

Map 3: Total solar eclipse of 21 June 2001 – the eclipse path through

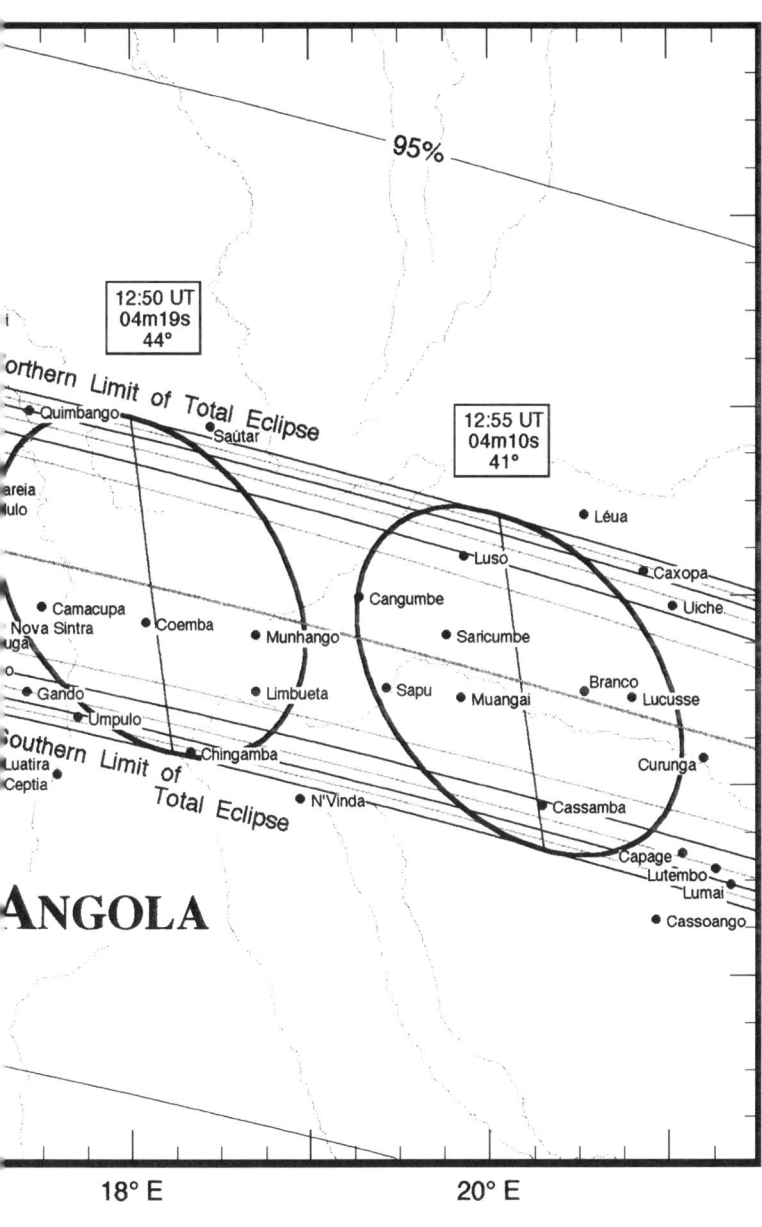

95%

12:50 UT
04m19s
44°

12:55 UT
04m10s
41°

orthern Limit of Total Eclipse

●Quimbango
●Saútar

areia
Iulo

●Léua

●Luso

●Caxopa

●Cangumbe

●Uiche

●Camacupa
Nova Sintra   ●Coemba
uga

●Munhango

●Saricumbe

o
●Gando

●Limbueta

●Sapu   ●Muangai

●Branco
●Lucusse

●Umpulo

outhern Limit of

●Chingamba

Luatira
Ceptia

Total Eclipse

●N'Vinda

●Curunga

●Cassamba

Capage●
Lutembo
Lumai

ANGOLA

●Cassoango

18° E   20° E

Angola.

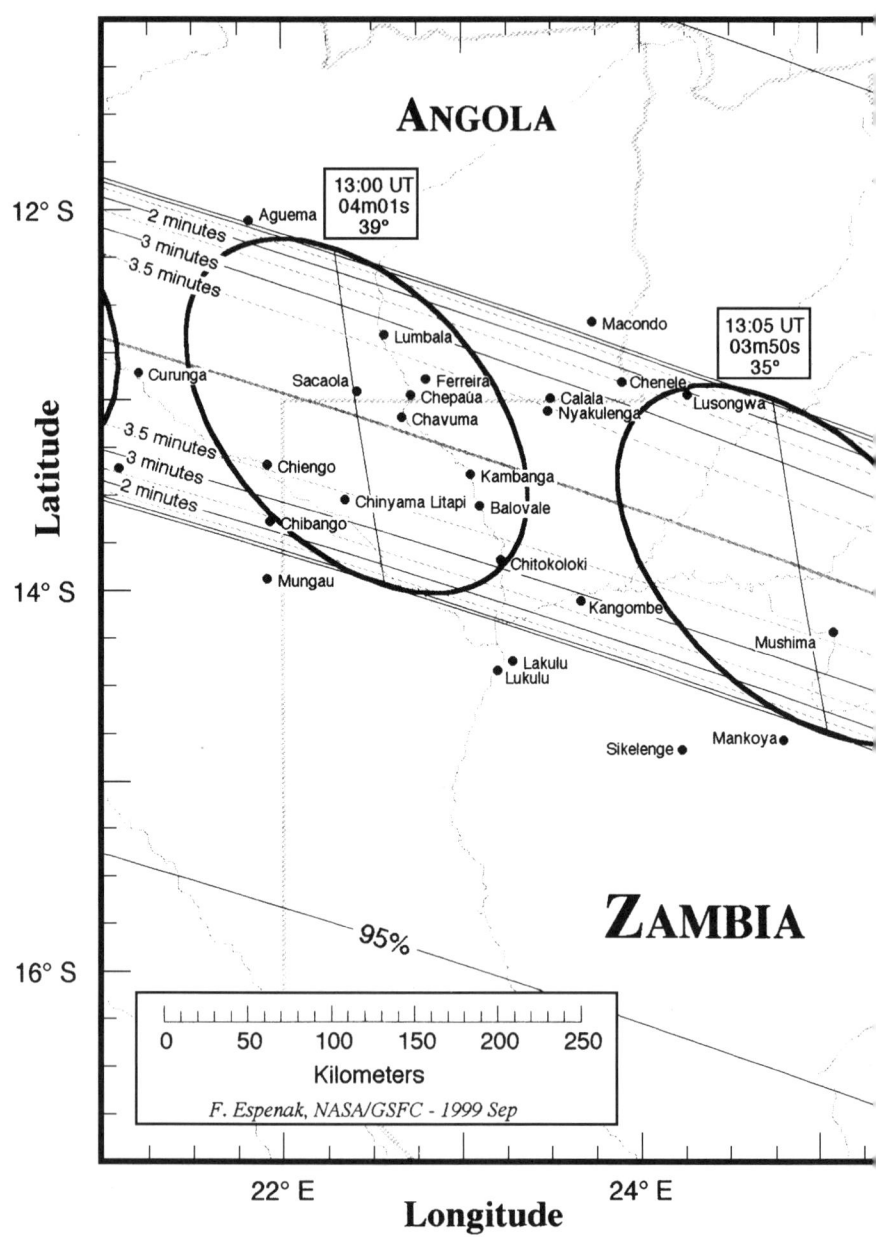

Map 4: Total solar eclipse of 21 June 2001 – the eclipse path through

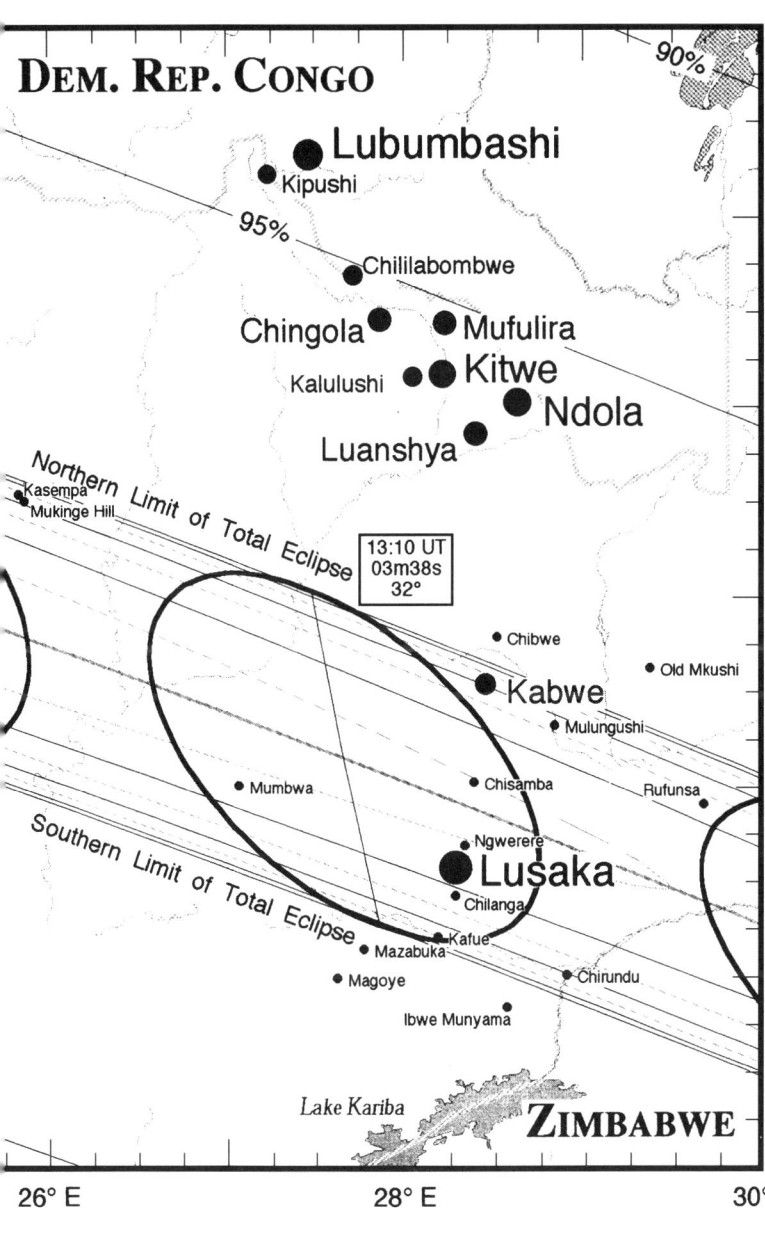

DEM. REP. CONGO

90%

● Lubumbashi

● Kipushi

95%

● Chililabombwe

Chingola ● ● Mufulira

Kalulushi ● ● Kitwe

● Ndola

Luanshya ●

Northern Limit of Total Eclipse

● Kasempa
● Mukinge Hill

13:10 UT
03m38s
32°

● Chibwe

● Old Mkushi

● Kabwe

● Mulungushi

● Mumbwa

● Chisamba

Rufunsa ●

Southern Limit of Total Eclipse

● Ngwerere

● Lusaka

● Chilanga

● Kafue

● Mazabuka

● Chirundu

● Magoye

● Ibwe Munyama

Lake Kariba

ZIMBABWE

26° E   28° E   30°

Zambia.

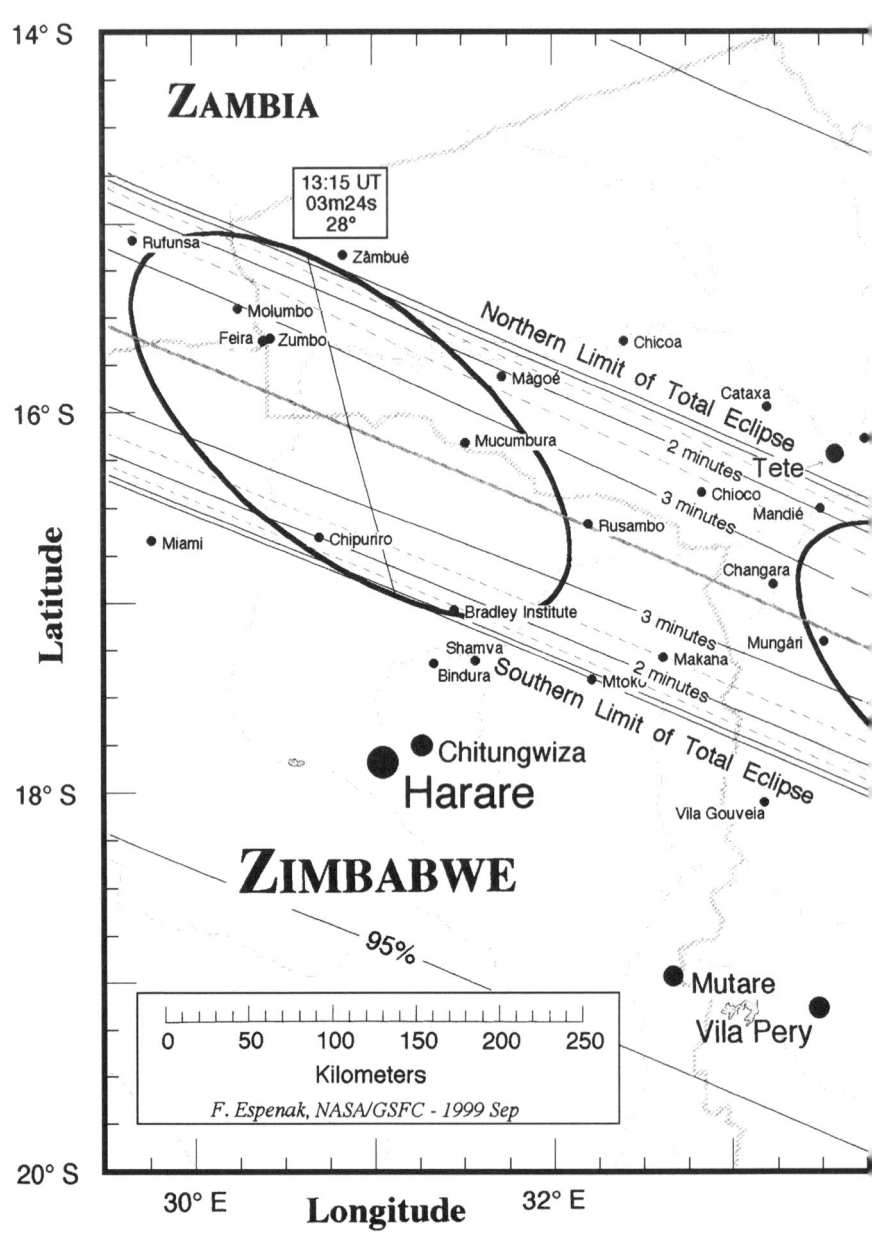

Map 5: Total solar eclipse of 21 June 2001 – the eclipse path through

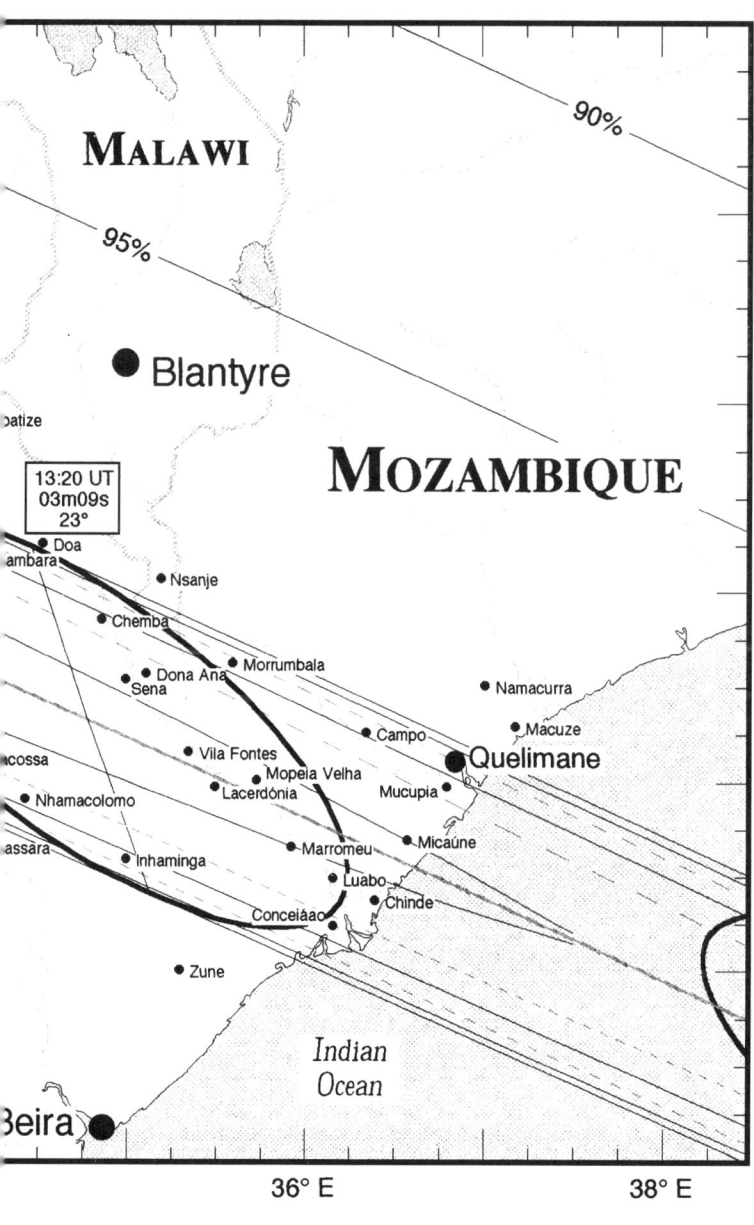

MALAWI

90%

95%

● Blantyre

atize

MOZAMBIQUE

13:20 UT
03m09s
23°

● Doa
ambara

● Nsanje

● Chemba

● Morrumbala

● Dona Ana
Sena

● Namacurra

● Campo

● Macuze

● Quelimane

cossa

● Vila Fontes

● Mopeia Velha
Lacerdónia

Mucupia ●

● Nhamacolomo

assara

● Inhaminga

● Marromeu

● Micaúne

● Luabo

Conceiâao

Chinde

● Zune

Indian
Ocean

Beira ●

36° E

38° E

Zimbabwe and Mozambique.

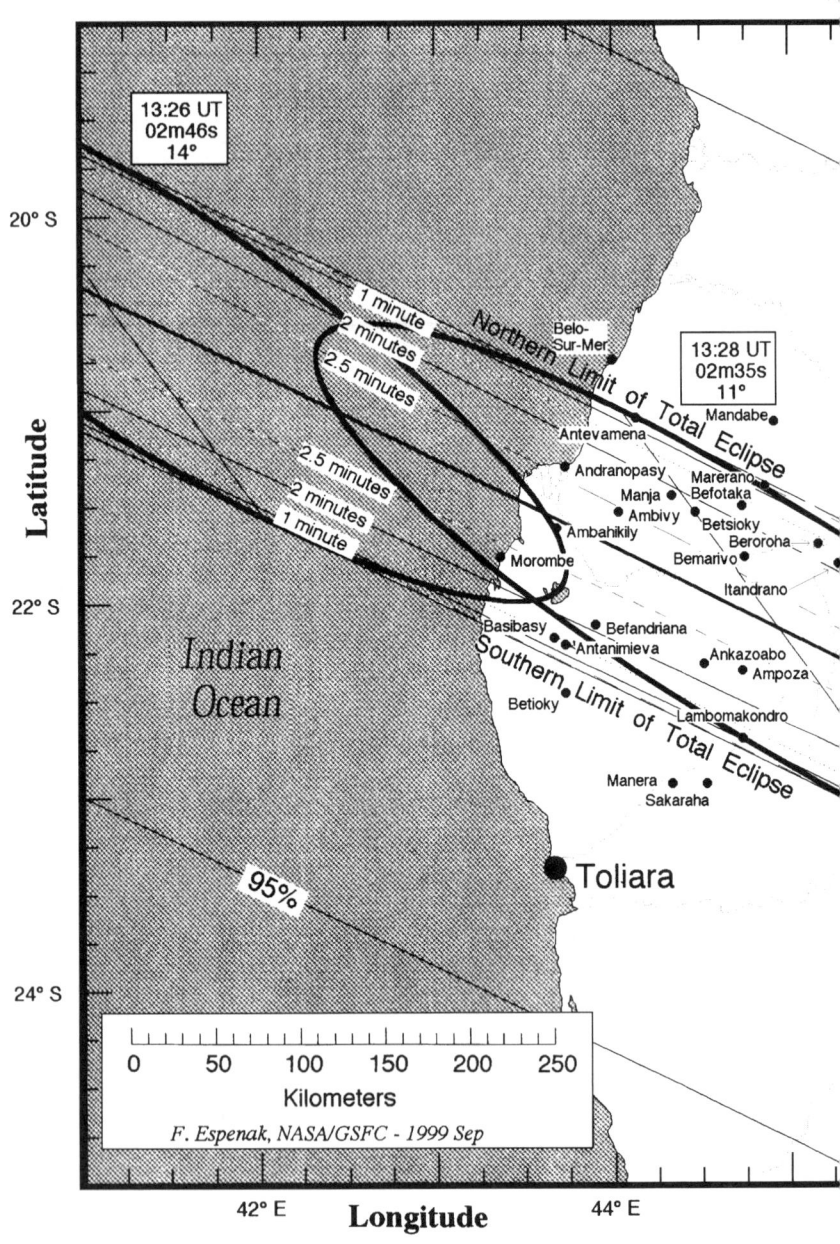

13:26 UT
02m46s
14°

20° S

Latitude

22° S

24° S

1 minute
2 minutes
2.5 minutes

Northern Limit of Total Eclipse

Belo-Sur-Mer

13:28 UT
02m35s
11°

Mandabe

Antevamena
Andranopasy
Manja
Ambivy
Ambahikily
Morombe

Marerano
Befotaka
Betsioky
Beroroha
Bemarivo
Itandrano

2.5 minutes
2 minutes
1 minute

Indian
Ocean

Basibasy
Betioky

Befandriana
Antanimieva
Southern Limit of Total Eclipse

Ankazoabo
Ampoza
Lambomakondro

Manera
Sakaraha

95%

Toliara

0   50   100   150   200   250
Kilometers

F. Espenak, NASA/GSFC - 1999 Sep

42° E   **Longitude**   44° E

Map 6: Total solar eclipse of 21 June 2001 – the eclipse path through

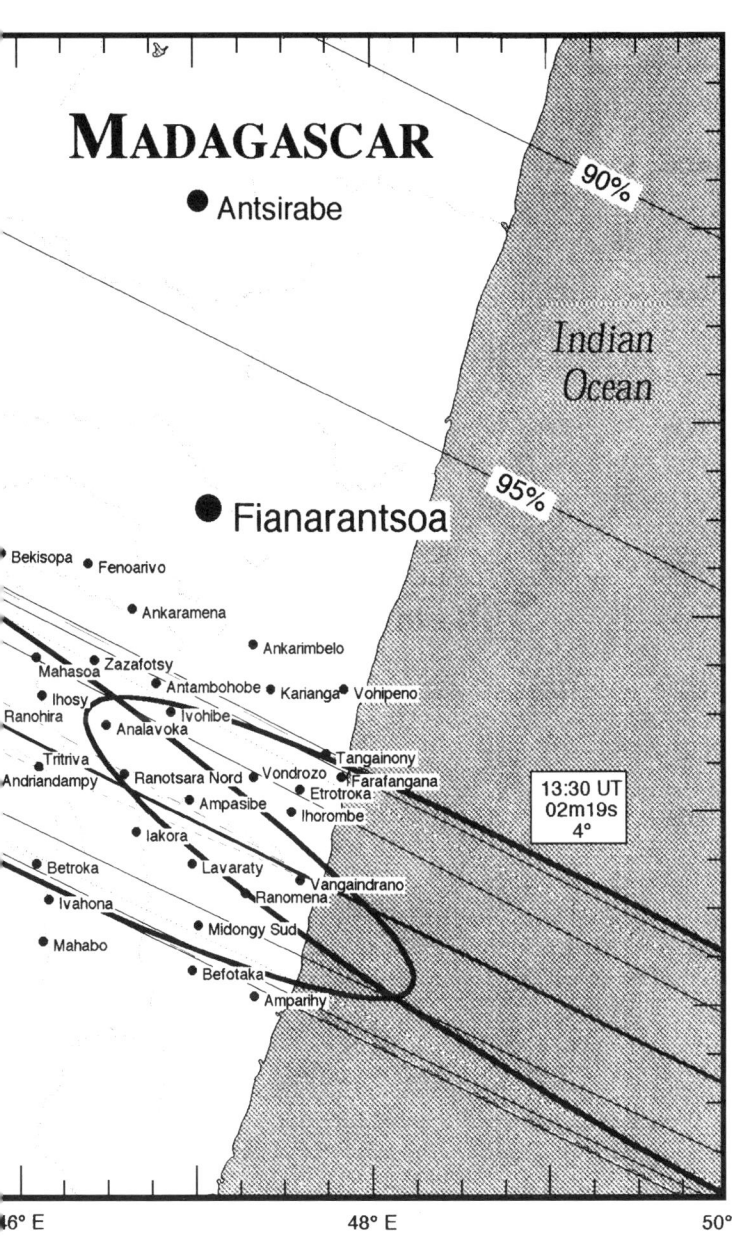

Madagascar.

Eclipse of 4 December 2002

Overview

In December 2002 another total eclipse will cross Southern Africa, this time a little further south than the one in June 2001. This slightly more southerly passage may tempt some people to pass up the June 2001 eclipse in favour of this later one. However there are a number of discouraging factors, relating mainly to the path running through relatively inaccessible parts of the continent.

This time, the track of totality runs through a much less populated area than the earlier solar eclipse, with fewer large or medium-sized towns and correspondingly fewer tourist facilities. It will be almost midsummer and much of the eclipse runs through the summer rainfall area with its attendant rainclouds. Fortunately, this factor may not have significant impact on viewing conditions, because the eclipse occurs very early in the morning and the rain usually occurs in the afternoons.

The period of totality is very short compared with the 2001 eclipse, so there will not be much time to take photographs or even just to take in the wonder of the spectacle. Totality drops to under a minute as soon as one strays a little off the centreline.

The average daytime temperatures along the entire path of totality across Africa range between 19°C and 26°C. In most places the sun is 15° or more above the horizon when the eclipse starts so it will already be quite warm. By the time of totality between 06h00 and 06h30 (Universal Time) the Sun will have risen to between 20° and 45° above the horizon as the shadow moves from west to east across the continent. (To correct Universal Time to local time, see the notes at the head of tables 3 and 4a to 4d.) This will be a comfortable angle both for looking at the Sun and photographing it. The eclipse will be above the low-angle atmospheric haze which is often apparent soon after sunrise in hot climates.

Angola

Once again, this is not likely to be a popular place for eclipse observation in view of the internal strife, but also because the length of totality will be under a minute for the more populated western regions. Average rainfall at Huambo, recorded over several years, is 220 mm per month for November and December. Attempts to obtain records of the number of hours of sunshine during the same period were not successful. Further inland there are very few towns of any size near the centreline and the difficult accessibility, as well as the ongoing civil war, will deter most visitors.

Table 3: Local details of the total eclipse of 4 December 2002

NB: All times are quoted in Universal time (previously GMT) and should be corrected to local time by adding the following: Angola: 1 hour; Zambia, South Africa & Zimbabwe: 2 hours; Mozambique: 3 hours

Site	1st Contact h m s	Alti-tude	2nd Contact h m s	3rd Contact h m s	4th Contact h m s	Alti-tude	Totality duration
ANGOLA							
Benguela	04:59:00	8	05:57:18	05:58:05	07:03:37	36	0m 47s
Chitembo	04:59:49	9	05:58:43	05:59:15	07:05:34	38	0m 32s
Ecunha	04:59:11	7	05:57:24	05:57:56	07:03:20	36	0m 32s
Huambo	04:59:08	8	05:57:23	05:58:16	07:03:46	36	0m 52s
Rivungo	05:02:37	15	06:04:18	06:04:54	07:14:50	46	0m 36s
ZAMBIA							
Sinjembela	05:03:50	17	06:05:55	06:07:02	07:17:31	47	1m 08s
CAPRIVI							
Kongola	05:04:24	17	06:06:47	06:07:44	07:18:35	48	0m 56s
BOTSWANA							
Ngwezumba	05:05:18	19	06:08:24	06:09:31	07:21:18	50	1m 07s
ZIMBABWE							
Beit Bridge	05:11:35	26	06:17:51	06:19:12	07:34:50	59	1m 21s
Plumtree	05:08:39	23	06:13:32	06:14:49	07:28:49	55	1m 17s
SOUTH AFRICA							
Messina	05:11:48	26	06:18:10	06:19:24	07:35:08	59	1m 14s
Punda Maria	05:12:38	27	06:19:33	06:20:52	07:37:17	60	1m 19s
Shingwedzi	05:13:23	28	06:20:33	06:21:59	07:38:39	61	1m 26s
MOZAMBIQUE							
Canicado	05:16:03	30	06:24:20	06:25:42	07:43:35	64	1m 22s
Mabalane	05:14:58	30	06:22:54	06:24:19	07:41:52	63	1m 25s
Xai-Xai	05:17:16	31	06:26:08	06:27:13	07:45:43	65	1m 05s
AUSTRALIA							
Streaky Bay	08:10:24	20	09:10:10	09:10:43	10:05:09	-2	0m 32s

Table 3 details the path of totality for the eclipse of 4 December 2002, its duration and the predicted times of the various contacts. Partial eclipse times are given in tables 4a-d. The maps on page 52 show the geographical path of totality.

Zambia

Only a narrow strip on the south-western border falls within the northern half of totality and there is little habitation. Half of the Sioma Ngwezi National Park is in the path of totality. Other than this there is good accommodation at Livingstone and the Victoria Falls nearby where there will be a partial eclipse of over 99%. Just across the border into Botswana or Caprivi there are also good viewing opportunities. Records taken at Lusaka indicate the likelihood of around 7 hours of sunshine per day at this time of year, but this is some 500 km from the line of totality. It is hoped that conditions will not vary too much over this distance.

Table 4a: Local details of the partial eclipse of 4 December 2002

NB: All times are quoted in Universal time (previously GMT) and should be corrected to local time by adding the following: Angola: 1 hour; Zambia, South Africa & Zimbabwe: 2 hours; Mozambique: 3 hours

Site	1st Contact h m s	Alti- tude	Max eclipse h m s	Alti- tude	Max percent obscured	4th Contact h m s	Alti- tude
ANGOLA (see also table 3 for totality)							
Chibango	04:59:38	14	06:01:02	28	92,6	07:10:37	44
Conda	04:57:51	5	05:55:43	19	98,9	07:00:35	34
Lobito	04:59:13	5	05:56:57	19	98,4	07:01:35	34
Luanda	04:56:05	3	05:53:08	16	93,6	06:56:57	31
Porto Amboim	04:57:38	5	05:55:13	18	98,8	06:59:42	33
BOTSWANA (See also table 3 for totality)							
Francistown	05:09:32	23	06:15:02	38	99,3	07:29:35	55
Gabarone	05:14:41	24	06:19:21	38	86,0	07:32:30	55
CONGO							
Brazzaville	04:52:49	3	05:49:04	16	75,2	06:52:00	30
Pointe-Noire	04:53:41	0	05:49:17	13	82,9	06:51:15	27
DEMOCRATIC REPUBLIC OF CONGO							
Kinshasa	04:52:50	3	05:49:06	16	75,3	06:52:04	30
Kolwezi	04:57:27	15	05:59:07	30	76,5	07:09:15	46
Lubumbashi	04:58:46	18	06:01:28	32	75,4	07:12:58	49
GABON							
Libreville	04:52:17	5	05:45:36	8	70,8	06:44:36	21

Table 4b: Local details of the partial eclipse of 4 December 2002

NB: All times are quoted in Universal time (previously GMT) and should be corrected to local time by adding the following: Angola: 1 hour; Zambia, South Africa & Zimbabwe: 2 hours; Mozambique: 3 hours

Site	1st Contact h m s	Alti-tude	Max eclipse h m s	Alti-tude	Max percent obscured	4th Contact h m s	Alti-tude
LESOTHO							
Maseru	05:23:02	28	06:28:17	42	76,2	07:41:39	58
MALAWI							
Blantyre	05:06:06	28	06:13:15	43	72,1	07:30:24	61
Lilongwe	05:03:37	25	06:09:23	41	69,3	07:24:52	58
MOZAMBIQUE (See also table 3 for totality)							
Beira	05:10:31	30	06:19:23	45	84,6	07:38:22	64
Changara	05:06:12	26	06:13:14	42	79,3	07:30:09	60
Chinde	05:10:04	31	06:19:07	47	77,7	07:38:26	65
Maputo	05:18:17	31	06:27:07	46	96,1	07:45:20	64
Quelimane	05:09:34	31	06:18:27	47	74,4	07:37:37	65
Tete	05:05:40	26	06:12:32	42	76,4	07:29:17	60
MADAGASCAR							
Ambahikily	05:20:05	40	06:33:33	57	71,0	07:57:37	77
Andranopas	05:19:46	40	06:33:05	57	69,9	07:57:01	76
Antananarivo	05:22:19	44	06:35:10	61	54,5	07:58:25	80
Farafangana	05:25:41	45	06:41:09	62	67,6	08:07:02	82
Fianarantsoa	05:24:08	44	06:38:45	61	63,2	08:03:51	81
Toliara	05:22:15	41	06:36:32	58	76,5	08:01:20	77
NAMIBIA							
Windhoek	05:11:37	15	06:11:04	28	75,1	07:17:37	43

Caprivi

The eastern bulge of Caprivi lies right on the eclipse path and good roads run through this area. The nearest location where rainfall records could be traced was at Livingstone where the monthly average for November was 78 mm rising to 176 mm in December.

Botswana

The Chobe National Park lies right across the track of the eclipse and could prove a pleasant place for a combined game holiday and view of the eclipse. South-east of this park is a fairly sparsely populated area, although a good road runs lengthwise through it from Kazangula on the north side to Francistown in the south. The distance between these towns is 530 km and it would require a very early start to be about half-way by eclipse time. Fortunately there are more convenient places nearer to these towns which are both less than 50 km from the edge of totality.

Francistown has the lowest average rainfall of all the records found for places along the route, with 59 mm in November and 126 mm in December, but no records of hours of sunshine could be found. The road north towards Bulawayo crosses the Zimbabwe border just before the centreline of totality.

Zimbabwe

The shadow path enters the country in North Matabeleland where the Libuti Camp of Hwange National Park lies right on the edge of totality. An early morning drive 30 km to the south should provide a memorable experience. Bulawayo is some 50 km outside the path of totality which is easily accessible on the road towards the Botswana border and Francistown. Table 3 gives predictions for Plumtree, which is a short distance inside the border. Rainfall during November and December is around 100 mm per month, which should leave the mornings relatively clear. Bulawayo has recorded an average of 7,8 hours of sunshine per day in November/December over a four-year period. Transport drivers waiting for clearance to cross into South Africa at Beit Bridge will get a near centreline view of totality shortly after 8.15 am local time.

South Africa

Only a tiny corner of South Africa from Messina into the northern part of the Kruger Park is in line with the path of totality. There are several camps in that part of the Park and the one at Shingwedzi is right on the centreline of the eclipse with 1m 26s of totality (see table 3 on page 45). This looks to be the prime prospect for those who like to see wildlife and are also keen to do some eclipse watching.

It is a pity that totality misses all the bigger centres with even Thohoyandou being just outside the path, although it does see a 99.9% partial event.

Table 4c: Local details of the partial eclipse of 4 December 2002

NB: All times are quoted in Universal time (p eviously GMT) and should be corrected to local time by adding the following: Angola: 1 hour; Zambia, Sout i Africa & Zimbabwe: 2 hours; Mozambique: 3 hours

Site	1st Contact h m s	Alti-tude	Max eclipse h m s	Alti-tude	Max percent obscured	4th Contact h m s	Alti-tude
SOUTH AFRICA (See also table 3 for totality)							
Alexandria	05:31:00	30	06:34:26	43	63,3	07:45:00	57
Benoni	05:17:39	27	06:23:53	42	86,7	07:38:49	58
Bloemfontein	05:22:23	27	06:26:44	41	74,2	07:38:59	56
Boksburg	05:17:25	27	06:23:30	41	86,6	07:38:17	58
Cape Town	05:31:48	23	06:29:12	35	49,6	07:32:24	48
Carletonville	05:17:34	26	06:23:06	41	84,3	07:37:09	57
Daveyton	05:17:22	27	06:23:34	42	87,1	07:38:31	58
Durban	05:24:34	31	06:32:09	46	81,6	07:48:15	62
East London	05:29:51	31	06:34:37	44	67,7	07:46:52	59
Evaton	05:17:53	27	06:23:45	41	85,0	07:38:12	58
Germiston	05:17:26	27	06:23:28	41	86,4	07:38:12	58
Johannesburg	05:17:26	27	06:23:22	41	85,9	07:37:58	58
Kempton Park	05:17:14	27	06:23:19	41	86,8	07:38:07	58
Kimberley	05:21:24	25	06:24:54	39	72,9	07:36:09	55
Klerksdorp	05:18:28	26	06:23:29	40	81,3	07:36:49	56
Mamelodi	05:16:41	27	06:22:49	41	87,9	07:37:44	58
Mdantsana	05:29:41	30	06:34:19	44	67,5	07:46:24	59
Natalspruit	05:17:36	27	06:23:37	41	86,1	07:38:19	58
Pietermaritzburg	05:23:51	31	06:31:00	45	81,1	07:46:39	62
Port Elizabeth	05:31:36	29	06:34:23	42	61,2	07:44:06	56
Pretoria	05:16:38	27	06:22:42	41	87,7	07:37:30	58
Soweto	05:17:24	27	06:23:16	41	85,8	07:37:47	58
Thohoyandou	05:12:55	27	06:20:15	42	99,9	07:36:58	60
Umtata	05:27:13	30	06:32:58	44	73,0	07:46:38	60
Vereeniging	05:18:05	27	06:23:58	41	84,8	07:38:26	58

Table 4d: Local details of the partial eclipse of 4 December 2002

NB: All times are quoted in Universal time (previously GMT) and should be corrected to local time by adding the following: Angola: 1 hour; Zambia, South Africa & Zimbabwe: 2 hours; Mozambique: 3 hours

Site	1st Contact h m s	Alti-tude	Max eclipse h m s	Alti-tude	Max percent obscured	4th Contact h m s	Alti-tude
SWAZILAND							
Mbabane	05:18:19	30	06:26:13	45	92,1	07:43:16	62
ZAMBIA (See also table 3 for totality)							
Chavuma	04:59:08	14	06:00:40	28	89,5	07:10:29	44
Chingola	04:59:37	19	06:02:50	33	77,2	07:14:59	50
Chisamba	05:02:06	20	06:06:29	35	83,7	07:20:02	52
Kabwe	05:01:36	20	06:05:49	35	81,9	07:19:10	52
Kafue	05:02:55	21	06:07:29	36	86,6	07:21:15	53
Kambanga	04:59:28	14	06:01:14	29	89,7	07:11:21	45
Kitwe	04:59:59	19	06:03:26	34	77,4	07:15:54	51
Livingstone	05:04:49	20	06:08:55	34	97,5	07:21:53	51
Luanshya	05:00:20	20	06:03:59	34	78,0	07:16:41	51
Lusaka	05:02:32	21	06:07:01	35	85,3	07:20:42	53
Mufulira	04:59:46	19	06:03:06	34	76,5	07:15:25	50
Mukinge Hill	04:59:58	17	06:02:51	32	84,3	07:14:29	48
Mumbwa	05:01:45	19	06:05:37	34	86,5	07:18:28	51
Mushima	05:00:33	17	06:03:23	31	88,2	07:14:52	48
Ndola	05:00:16	20	06:03:55	34	77,0	07:16:39	51
Rufunsa	05:02:40	22	06:07:35	37	81,3	07:21:50	54
ZIMBABWE (See also table 3 for totality)							
Bindura	05:05:37	24	06:11:55	40	85,3	07:27:50	57
Bulawayo	05:08:19	24	06:14:12	39	98,8	07:29:20	56
Harare	05:06:12	25	06:12:41	40	86,8	07:28:47	58

Mozambique

Rainfall figures for November/December of around 85 mm per month indicate probable good viewing. The records of sunhine show daily averages of 8,1 hours and 7,4 hours at two places along the coast. Records made 80 km inland at Chokwe near Canicado, well within the path of totality, show an average of 7,5 hours per day. The eclipse is well above the horizon at 30 to 65° elevation and totality lasts more than a minute at a comfortable 45°.

Australia

This is included to illustrate two things. Firstly, it shows that eclipses can cover huge distances on Earth, and secondly, it demonstrates the speed at which the shadow travels. Leaving the coast of Mozambique at 6h27 UT, it comes ashore at Streaky Bay in the Great Australian Bight at 9h10 UT. That's nearly one-third of the circumference of the Earth in two-and- three-quarter hours. The Sun will set in the west by the time the eclipse ends in Australia, but it should be possible to get some fine photographs across the sea, especially if there are any yachts about.

Totality

The Moon will be further from the Earth than it was for the June 2001 eclipse, which will reduce the size of the shadow on the surface, and the maximum period of totality will be less than 1 minute 35 seconds. This does not allow for much time in totality and more serious observers will probably make their major expedition efforts in June 2001, leaving this one only as a backup in case of unexpected difficulties or bad weather. The Sun will be almost due east from near the South Africa/Zimbabwe border region and it should be possible to see Mercury to its lower right. Venus will be almost directly overhead with Mars quite close to it, but not so bright. Jupiter will be far to the west and will probably not be noticed by many. Figure 9 shows the positions of planets and a few bright stars.

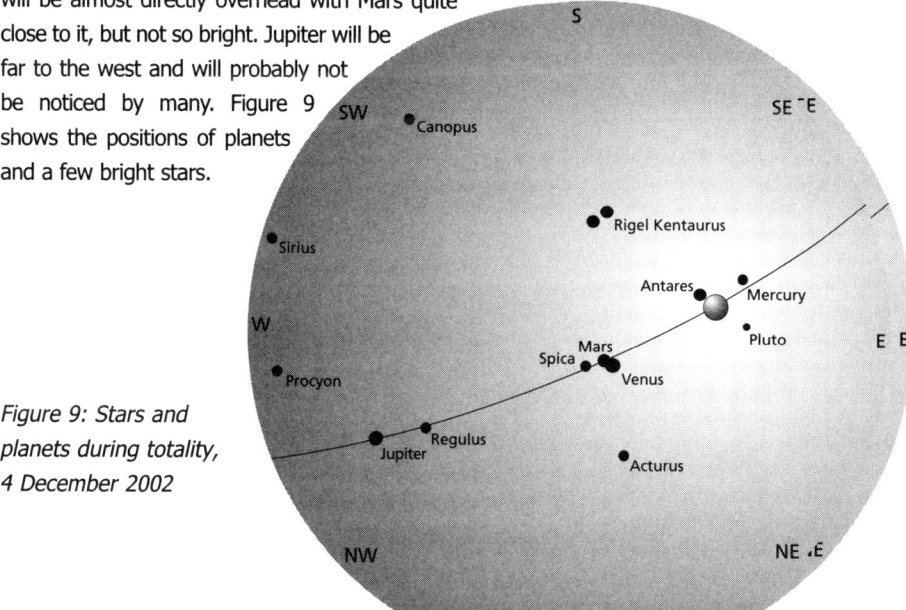

Figure 9: Stars and planets during totality, 4 December 2002

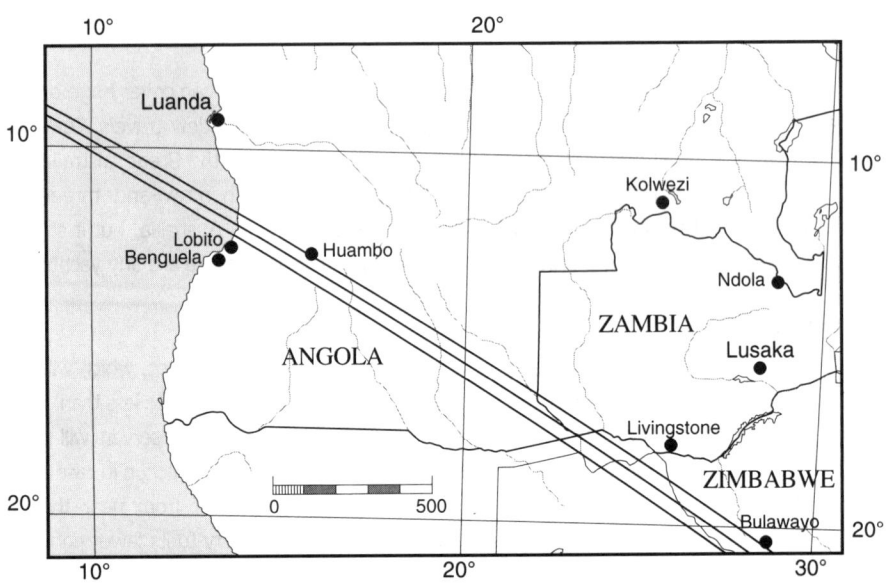

Map 7: Total solar eclipse of 4 December 2002 – eclipse path through Angola and Botswana

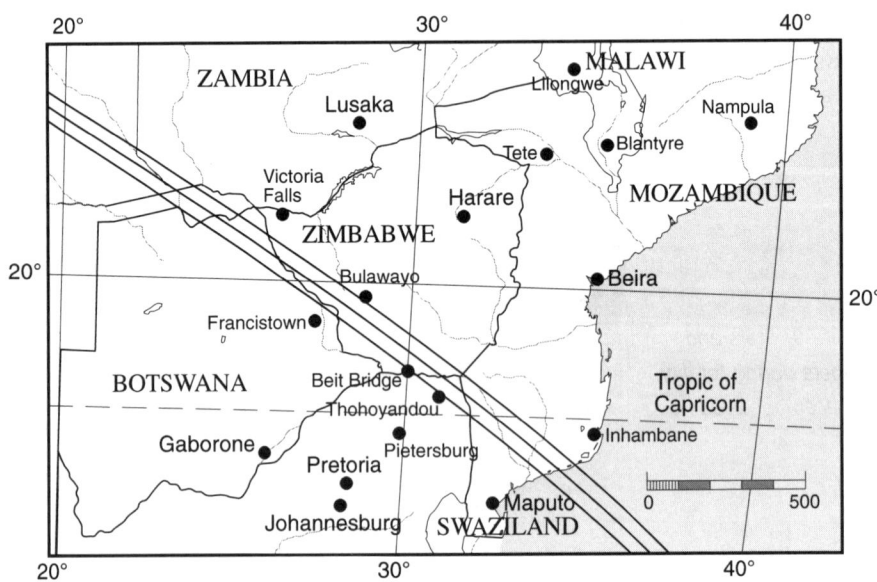

Map 8: Total solar eclipse of 4 December 2002 – eclipse path through Zimbabwe.

Photographing eclipses

Solar eclipses

Cameras and lenses

Almost any camera and film can be used to photograph eclipses but some have specific advantages and are more convenient than others. Single lens reflex cameras view the subject through the same lens that is used to take the photograph, and, of these, the most commonly available is the 35 mm camera. Other sizes are not so freely available and there may not be such a variety of films to choose from.

Twin lens reflex cameras, which were popular some years ago, make it easy to compose the picture. However, you would need a solar filter over both lenses. In order to see what you are doing during the few minutes of totality, you would need to remove the filter from both camera and finder – and put them both back again before the Sun starts to reappear. Clearly then, the only real choice is a single lens reflex.

Lenses should be interchangeable and you may even find it convenient to use two cameras with different focal length lenses to capture varying effects. The usual 50 mm lens will not show a very big image of the eclipsed Sun, but will be useful for photographs of fellow watchers. It also provides a wide field of view to include planets and bright stars which can usually be seen during totality. Longer focus lenses give a larger image size on the film, but bear in mind that at totality, the corona is likely to be at least twice the diameter of the Sun itself.

Table 5 shows the size of the solar image on 35 mm film when using various focal lengths. The actual size of the 35 mm negative or slide is 24 mm x 36 mm, so it is wise to calculate in advance the effect of the various lenses available. A quick rule of thumb to determine image size of the Sun (or of the Moon) is to divide the focal length of the lens by 110. A 200 mm lens will give a 200/110 = 1,81 mm diameter image on the film, while a 500 mm lens will increase this to 500/110 = 4,54 mm, which is just under one fifth of the 24 mm film width. Even allowing for the additional size of the corona, a 500 mm lens will record totality and allow sufficient space all around, even if the camera is not perfectly aimed with the Sun dead centre.

It is usually considered that the longest lens to use should be a 1 200 mm lens, but the author prefers to play it safe with a maximum length of 1 000 mm. Most of his eclipse pictures have been taken with a 600 mm lens, although he has experimented with up to 2 000 mm during the partial phases. It is possible to find focal converters of various strengths at photographic shops that will increase the focal length of any lens by a given factor. The

Table 5:

Sun's image size and field of view for lenses of various focal lengths.		
Focal length of lens in mm	Solar image diameter on film in mm	Field of view of negative in degrees
28	0,25	49 x 74
35	0,32	39 x 59
50	0,45	27 x 40
100	0,91	13 x 20
135	1,23	10 x 15
150	1,36	9 x 14
200	1,82	7 x 10
300	2,73	4,6 x 6,9
500	4,55	2,7 x 4,1
600	5,45	2,3 x 3,4
1 000	9,09	1,4 x 2,1
1 200	10,91	1,1 x 1,7

factor may be 1,5, 2,0 or even 3,0 times. Thus a 200 mm lens with a 3x converter will act like one of 600 mm, to give an image of 600/110 = 5,45 mm.

Zoom lenses can be used just as easily as those of fixed focus, but it may be worth while to do a few experiments in advance of the actual eclipse. Some zooms are a little loose and, when pointed upwards, they tend to turn by themselves, both between shots and even during an exposure. If you are worried about movement, whether it be of zoom or focus, don't hesitate to hold it in place with masking or packing tape.

Films

The first choice is between slides or prints. You will probably be better off with prints (negative film) unless you are reasonably experienced. This is because you can make errors with the exposure of negative film without its becoming a disaster. Over- or underexposure of four times (or one quarter) will not wreck your pictures – and you may even get away with more than this. Slide film demands greater respect and you run more risk of losing your record of the eclipse.

Cameras with automatic exposure settings will help to avoid problems with slide film exposures, but for some effects one deliberately wants to under- or overexpose the subject. A glance at the exposure tables will immediately show that a longer exposure is needed to bring out the corona further from the limb of the Sun. Similarly, shorter exposures are required to fully show prominences on the limb which would otherwise be lost in an overexposed part of the picture. You may wish to use both types of film in different cameras. In this case, be careful to match the lens to the type of film you plan to use for different shots.

Film speed denotes how much exposure is needed, with faster films requiring less exposure time. This is an important factor as the Sun moves across the sky and will make a blurred image if the exposure is too long. Many people do not realise that the Sun moves as much as its own diameter every 2 minutes. In truth, it is not the Sun that is moving at this speed, it is the Earth turning on its axis, giving the impression of the Sun, Moon, planets and stars all moving round the sky at the same speed. There are also individual motions of all those bodies but, from our point of view, they are very small when compared with the Earth's rotation. It is possible to correct for the turning of the Earth but that will be dealt with later.

The diameter of the Sun is a half degree or 30 minutes of arc and, as this is the distance moved in 2 minutes of time, it follows that a movement of 1 arc-minute takes 4 seconds of time. One second of time will produce a movement of only 1/120th of the Sun's diameter (equal to 15 arc-seconds) but this will be noticeable on pictures taken with a lens having a focal length over 300 mm. It is therefore wise to choose film that is reasonably fast.

Film speed is indicated by an ISO (ASA) number, with the larger numbers indicating faster film. The fastest films usually have a much coarser grain in their emulsion which can affect clarity, but ISO 400 films today have finer grain than even ISO 64 or 100 had a few years ago. The very fast ISO 1 600 and 3 200 films may show some grain, but ISO 200 or even ISO 400 should give no problems. All the major film manufacturers produce fine quality products and the choice is usually just personal preference.

For eclipse photography there should be no real problem with the maximum exposure times in Table 6 unless you have a very small aperture lens on your camera coupled with slow film, but these limits are very important if you later try to capture pictures of the night sky.

Tripods and Mountings

The slightest movement of the camera when taking pictures with a long focus lens will cause blur on the picture. To avoid this, it is necessary to use a tripod. Avoid light, flimsy models: test the tripod using a mounted camera with the longest lens you have available. Then, while looking through the viewfinder, give the tripod leg a sharp knock; the vibration should settle down quickly. If not, try a different tripod until you find the best possible option within your price range.

The adjustable legs should be kept as short as possible to enhance

Table 6:

Longest exposures to give sharp images of celestial objects without a drive	
Effective focal length in mm	Maximum exposure in secs
50 - 150	2
150 - 350	1
350 - 700	0,5
700 - 1 500	0,25

stability, but must be high enough for easy use when the photographer is seated comfortably. For astrophotography, it is a good idea to suspend a heavy weight underneath the tripod to hold it firmly in place and further reduce any possible vibration. Even the mirror in the camera flipping up for the picture to be taken and then returning again can cause a tremor that is recorded on the picture. This can be avoided completely if you are lucky enough to own a camera that enables the mirror to be locked up just prior to the exposure, as is possible with some Nikon, Canon and one model of Pentax.

Even pressing the shutter release button while the camera is on the tripod will cause a shake. In fact, the shakes caused by nervous photographers are probably the biggest single cause of spoiled pictures. Use a cable release whenever you use a tripod – and test it first to make sure it moves smoothly. These small items are easily lost and, as they are cheap, it is prudent to carry a spare.

It is simple to use a camera viewfinder under normal circumstances, but quite different when it is pointing up into the sky. Check the expected elevation angle of the eclipse and set up your equipment beforehand to sort out any difficult positions. If you have (or can borrow) an angled viewfinder, you will find it to be very useful.

Photo: Cliff Turk

Plate 9: A tripod is an important accessory.

Another accessory which the author has found to be useful is a piece of aluminium channel fixed across the top of the camera mount on the tripod. This channel is then drilled so that two cameras can be mounted on it, aligned to be centred on the same spot. In this way it is possible to aim both cameras through one angle finder, although taking pictures with one camera while looking through the viewfinder of its neighbour does puzzle onlookers.

Finally, there is the question of whether to use a clock drive on the camera mounting. The purpose of these drives is to counteract the rotation of the Earth by setting up the camera (or telescope) on a shaft that is placed exactly parallel to the Earth's axis. For long exposures of several minutes when taking pictures of the night sky, this is essential, but for our very short exposures of the Sun it is not necessary. Amateur astronomers who know about equatorial mountings on telescopes will have convenient access to all that is necessary, but otherwise it is probably best to stay away from such equipment.

Photo: Cliff Turk

Plate 10: Two cameras mounted side by side on a length of aluminium channel, both protected by Mylar filters.

Taking the Pictures

Types of filters for looking at or taking pictures of the Sun are dealt with in the chapter on 'Eclipse observing'. Cameras also require filters, or the intense rays of the Sun will burn the film. Normal photographic filters will NOT do the job. Specialized solar filters are available for most sizes of camera lenses, but have to be ordered from overseas and are quite expensive. The most easily available material in South Africa is Mylar which is used for packaging. It is a plastic sheet with a coating of aluminium that reflects most of the light instead of letting it through. Using two layers of Mylar held in front of the camera lens by an ultra-violet filter works well. Just make sure that no direct light at all can sneak through at the edges.

Mylar gives a blue image of the Sun, but this can be avoided by using an orange or yellow filter instead of ultra violet. In fact, the author uses a UV filter onto the camera lens with the orange filter in front and two layers of Mylar between them. The two filter holders are taped together so there is no chance of them coming apart and they are quick to remove for pictures during totality – and to replace when totality ends.

Don't worry if the mylar you obtain has a few wrinkles in it. Because Mylar is so thin, the amount of refraction caused by the light passing through a piece which is slightly angled to the lightpath is negligible. Wrinkled Mylar can be seen on the author's cameras in plate 8.

For many people, this will be their first sight of a solar eclipse of any kind and some will want to capture the very early stages when the Moon takes its first bite out of the Sun. The only way to do this is with a filter covering the lens of the camera and then most of us will have no way of knowing what exposure to give.

The only solution is to take a number of practice shots some weeks before the eclipse. Set up your equipment as you expect to use it on the great day, make sure the filter is firmly in place, aim at the Sun and with a camera aperture (stop) setting of f 5.6 take a series of exposures of 1/60 right down to 1/1 000 sec and if possible 1/2 000 sec. Then repeat the whole series with the stop down to f 16. Make a note of the order in which the exposures were taken so you can identify which is which after they are developed. Warn the processing shop what the pictures are so they can alter the density of prints to get solid black backgrounds instead of hazy grey ones as sometimes happens. You may even find you have some big sunspots on the trial shots. You should find that the sun looks darker at the limb than in the centre. This is because the Sun is made of burning gas and you do not see a hard surface, you are looking deep into it. Discuss which picture you think is best with the processor who will eventually also print your eclipse pictures.

Leave enough time to take another series of practice pictures just in case you have a complete disaster. When you have decided which of your pictures has the exposure you want, compare it with the exposure guide in table 7. Bear in mind that your filter may have a different

Table 7:

Exposure guide for solar eclipse photography			
a) Full or partial disc with 'Solar-Skreen' or ND 5.0 filter			
Focal ratio f	ISO (ASA) 100	ISO (ASA) 200	ISO (ASA) 400
2	1/2000	1/4000	--
2.8	1/1000	1/2000	1/4000
4	1/500	1/1000	1/2000
5.6	1/250	1/500	1/1000
8	1/125	1/250	1/500
11	1/60	1/125	1/250
16	1/30	1/60	1/125
22	1/15	1/30	1/60
b) During totality – no filter			
Subject & focal ratio f	ISO (ASA) 100	ISO (ASA) 200	ISO (ASA) 400
Baily's Beads			
5.6	1/1000	1/2000	1/4000
8	1/500	1/1000	1/2000
11	1/250	1/500	1/1000
Diamond Ring			
5.6	1/125	1/250	1/500
8	1/60	1/125	1/250
11	1/30	1/60	1/125
Prominences			
5.6	1/125	1/250	1/500
8	1/60	1/125	1/250
11	1/30	1/60	1/125
Inner corona			
2.8	1/60	1/125	1/250
4	1/30	1/60	1/125
5.6	1/15	1/30	1/60
Outer corona			
2.8	1/15	1/30	1/60
4	1/8	1/15	1/30
5.6	1/4	1/8	1/15

density from the standard 'Solar-Skreen' filter (a commercial product from the US) for which the guide is designed. Because the thickness and density of Mylar shows some variations and is often used double, the exposure table opposite (table 7) is for a standard density filter. Mylar users should take a few test shots with the Mylar in their possession, and will then be able to adjust the table exposures by one or two stops either way, as necessary. Once you have established the correct exposure for your pictures with one aperture and film, it is easy to adjust for different apertures and other film speeds.

Most of your pictures during the partial phases will need the same exposure, as the intensity of light from the solar surface is the same all over, except for that darker band round the limb. When the Sun gets down to a fine crescent you can increase the exposure by one or two stops because the source of light is becoming increasingly limited. The only other likely difference could be if there is any haze or even light cloud about. For this problem you will just have to make an intelligent guess – probably by opening the lens by one stop (lower number) to cope with the haze or cloud. Even after a careful test, it is still worthwhile increasing and decreasing the exposure (either with the stop or the time) each side of the predetermined exposure. You may never get another chance to photograph an eclipse so it is worth wasting some film to get at least one or two really good pictures – and you might get many.

Lunar eclipses

Cameras and Lenses
The advice regarding cameras and lenses for solar eclipes applies equally to lunar eclipses, with only one major exception. The image size is the same for both Sun and Moon, but the Moon has no corona around it. Two well-known makes of 200 mm diameter telescopes have focal lengths of 2 000 mm and can be confidently used for lunar photographs.

Film choice
Once again, the information pertaining to solar eclipses is relevant. Negative film allows more error than slides, but the latter is more useful for display to a group of people. The Moon actually moves across the sky slightly more slowly than the Sun, but for all practical purposes we can treat the speeds as the same. Exposure time limits to avoid movement showing in the picture for various focal lengths of lenses are the same as in table 6.

As the Moon reflects very much less light than the Sun radiates, it is not necessary to use any filters. In fact, they can be a nuisance as they are a further complication when trying to determine exposure times and apertures.

Because of the much greater range of available light for Moon pictures, it is probably better to use a film that is faster than would be used for solar photography. It is easier to reduce the exposure or use a smaller f ratio (higher number) to avoid overexposure than it is to use long exposures, because of the need to avoid detectable movement.

Tripods and Mountings

Details discussed above for solar eclipses are relevant to lunar eclipses as well. One advantage of lunar photography is that one does not have to travel to a specific very small path to capture totality. You may be able to find a nearby amateur astronomer who has an equatorially mounted telescope with a drive. If you can arrange to mount your camera piggy-back on his telescope, or better still to take photographs through it, you will have every facility you could wish for.

Taking the Pictures

Unlike the Sun, for which you use the same exposure for any stage of partial eclipse (except for the very last, finest crescent) the brightness of the partially-eclipsed Moon is constantly changing. Covering possible errors by doubling and halving your guesstimated exposures may not be enough. It is better to play safe and use half, quarter, double and quadruple exposures.

Here again, built-in exposure meters help, but remember they deal with average brightness of the whole picture and you have the problem of dark eclipsed portions and brighter sunlit parts. The contrast is often much more than expected. The meter takes into account the dark sky background as well as the image, so it will usually give too much exposure for any picture where the Moon does not fill the whole frame. Table 8 should give you a starting point, but you will have to experiment.

Many factors can affect the brightness of your subject and these include the transparency of the atmosphere on the particular day at your location, as well as the elevation angle at which you are working. Sometimes an eclipse just happens to be particularly dark (or bright) in relation to previous similar eclipses, and exposure estimates go awry. The very first picture the author ever took of the Moon came out almost perfectly – and he has been trying to equal that ever since!

To make matters worse, you cannot practice on (say) a first quarter Moon as a test for a 50 per cent eclipse. During an eclipse, the uneclipsed part will still be within the penumbra of the Earth's shadow and will be a lot less bright than the illuminated part at first quarter. At the same time, the eclipsed part will be brighter than the unlit part at first quarter. You can't win!

Table 8:

Exposure guide for lunar eclipse photography

(a) Partial eclipse – very rough guide only – can safely be reduced by factor of 30 for bright portions only

Focal ratio f	ISO (ASA) 100	ISO (ASA) 200	ISO (ASA) 400
2	1/8	1/15	1/30
2.8	1/4	1/8	1/15
4	1/2	1/4	1/8
5.6	2	1	1/4
8	5	2	1
11	12	5	2
16	25	12	5
22	60	25	12

(b) Total eclipse – fairly bright – can be increased by a factor of 20 for dark eclipses

Focal ratio f	ISO (ASA) 100	ISO (ASA) 200	ISO (ASA) 400
2	1	1/2	1/4
2.8	3	1	1/2
4	6	3	1
5.6	15	6	3
8	35	15	6
11	80	35	15
16	200	100	35

Theoretically, lunar eclipses should be visible from 50% of the Earth's surface at any time. However it must be borne in mind that for locations in high latitudes or close to 90° either side of the central longitudes (listed on page 63), the elevation of the eclipse will be very low in the sky. For practical purposes, the limit of visibility is probably no more than about 65° either side of the central longitude.

GLOSSARY

Annular eclipse: An eclipse where the shadow spot of the umbra does not quite reach the Earth, so that the Moon's image against the Sun is surrounded by a ring of visible light.

Centreline: The imaginary line on the Earth's surface along which the path of totality of an eclipse is centred.

Corona: The outer atmosphere and surface flares of the Sun, visible during a total eclipse.

Eclipse: Interception of the light of one heavenly body (Sun, Moon) by another, as seen from Earth.

Lunar eclipse: An eclipse of the Moon, where the Earth comes between the Sun and the Moon, and casts a shadow on the Moon.

Partial eclipse: Partial concealment of one heavenly body by another.

Partial stages: Those stages preceding and following a total eclipse.

Penumbra: Partial shadow surrounding the umbra, caused by the partial obstruction of the light from an extended object (e.g. the Sun).

Pinhole effect: The formation of an inverted image on a screen, by light passing through a pinhole in an opaque barrier.

Shadow spot: The shaded area caused by the umbra meeting the Earth.

Solar eclipse: An eclipse of the Sun, where the Moon comes between the Earth and the Sun, and casts a shadow on the Earth.

Total lunar eclipse: An eclipse where the Earth's shadow entirely covers the Moon.

Total solar eclipse: An eclipse where the Moon entirely covers the Sun.

Totality: That stage of a total eclipse when direct sunlight is entirely blocked, whether on the Earth or the Moon.

Umbra: Cone of full shadow where a light source is completely obscured by an intervening body.

ECLIPSES 2001 TO 2010

SOLAR ECLIPSES

Date	Type	Where Visible
2001 Jun 21	Total	Atlantic Ocean, Southern Africa, Madagascar
2001 Dec 14	Annular	Pacific Ocean, Central America
2002 Jun 10	Annular	Pacific Ocean
2002 Dec 4	Total	Southern Africa, Indian Ocean, Australia
2003 May 31	Annular	Iceland
2003 Nov 23	Total	Antarctica
2005 Apr 8	Ann/Total	Pacific Ocean, Central America, north of South America
2005 Oct 3	Annular	Atlantic Ocean, Spain, Africa, Indian Ocean
2006 Mar 29	Total	Atlantic Ocean, Africa, Turkey, Central & North Asia
2006 Sep 22	Annular	NE of South America, Atlantic Ocean, South Indian Ocean
2008 Feb 7	Annular	South Pacific Ocean, Antarctica
2008 Aug 1	Total	North Canada, Arctic Ocean, China, Siberia
2009 Jan 26	Annular	South Atlantic Ocean, Indian Ocean, Sumatra, Borneo
2009 Jul 22	Total	Asia, Pacific Ocean
2010 Jan 15	Annular	Africa, Indian Ocean, South and South East Asia
2010 Jul 11	Total	Pacific Ocean, far south of South America

LUNAR ECLIPSES (Penumbral only eclipses omitted)

Date	Type	Centre of area of visibility (Longitude)
2001 Jan 9	Total	55 E
2001 Jul 5	Partial	135 E
2003 May 16	Total	55 W
2003 Nov 9	Total	20 W
2004 May 4	Total	50 E
2004 Oct 28	Total	45 W
2005 Oct 17	Partial	180 E/W
2006 Sep 7	Partial	75 E
2007 Mar 3	Total	10 E
2007 Aug 28	Total	160 W
2008 Feb 21	Total	50 W
2009 Dec 31	Partial	70 E
2010 Jun 26	Partial	175 W
2010 Dec 21	Total	125 W

CONTACTS & ADDRESSES

This list is published to assist the reader to obtain information. It is not exhaustive and does not imply endorsement of the products or services listed.

Astronomy information (general)
•*Astronomical Handbook for Southern Africa.*
Published annually by Astronomical Society of Southern Africa, in December for following year.
R20 (incl postage) from: Business Manager ASSA, PO Box 9, Observatory 7935

Astronomy societies
•Astronomical Society of Southern Africa:
PO Box 9, Observatory, 7935
Website being revised & republished.
•Durban Centre, ASSA:
http://www.astronomical.lia.net
•Harare Centre, ASSA:
http://www.geocities.com/zimastro/assa.html
•Johannesburg Centre, ASSA:
http://www.aqua.co.za/assa_jhb/assa001q.htm
•Natal Midlands Centre, ASSA:
http://www.botany.unp.ac.za/nmc/nmc.htm
•Pretoria Centre, ASSA:
http//mafadi.aero.csir.co.za/assa/index.html
•Port Elizabeth Peoples Observatory Society:
PO Box 7988, Port Elizabeth 6055

Observatories
•Cederberg Observatory:
http://cw.scouting.org.za/obs
•South African Astronomical Observatory:
PO Box 9, Observatory, 7935
Tel: (021) 447 0025
www:http://www.saao.ac.za
E-mail: *director@saao.ac.za*

Planetaria
•Johannesburg Planetarium:
Yale Road, Milner Park
Tel: (011) 717 1392 Fax: (011) 339 2926)
•The Planetarium, SA Museum:
25 Queen Victoria St, Cape Town, 8000
Tel: (021) 424 3330

http://www.museums.org.za/sam/planet/planetar.htm

Eclipse information
•*http://sunearth.gsfc.nasa.gov/eclipse/TSE2001/TSE2001.html*
•*http://umbra.nascom.nasa.gov/eclipse/predictions/y-m-d.html*
NB For 'y-m-d' type '2001-June-21' or '2002-December-04'
•Harare Centre, ASSA:
E-mail: *zimbabweeclipse@hotmail.com*

Photography books
•Arnold, HJP, *Astrophotography: An Introduction.* Geo Philip Ltd, UK
•Covington, Michael, *Astrophotography for the Amateur.* Cambridge University Press
•Pasachoff, Jay M & Covington, Michael A, *The Cambridge Eclipse Photography Guide.* Cambridge University Press

Travel information
•Zambian eclipse website:
http://homepages.go.com/homepages/e/l/a/elampi/index.html
•Zimbabwe Tourism Authority:
http://www.icon.co.za/zta OR
http://www/zimtravel.com

Solar filter sources
•Celestron International,
2835 Columbia St, Torrance CA 90503, USA
•Meade Instruments Corp,
16542 Millikan Ave, Irvine CA 92714, USA
•Rainbow Symphony Inc,
6860 Canby Ave, #129, Reseda CA 91335, USA
•Thousand Oaks Optical,
Box 5044-289, Thousand Oaks CA 91359, USA
•Cliff Turk
E-mail: *cliffturk@yebo.co.za*